Routledge Revivals

Trends in Energy Use in Industrial Societies

Taken from a report for the Electric Power Research Institute, Joy Dunkerley's study aims to clarify the relationship between energy consumption and economic output in industrialised countries. Originally published in 1980 and using data from 1972, this study uses cross-country comparisons of energy use to stress the importance of new supply options and measures of controlling energy use without affecting living standards whilst also discussing the impact of energy prices and economic growth in the countries studied. This title will be of interest to students of environmental studies.

Trends in Energy Use in Industrial Societies

An Overview

Joy Dunkerley

First published in 1980
by Resources for the Future, Inc.

This edition first published in 2016 by Routledge
2 Park Square, Milton Park, Abingdon, Oxon, OX14 4RN
and by Routledge
711 Third Avenue, New York, NY 10017

Routledge is an imprint of the Taylor & Francis Group, an informa business

Publisher's Note
The publisher has gone to great lengths to ensure the quality of this reprint but points out that some imperfections in the original copies may be apparent.

Disclaimer
The publisher has made every effort to trace copyright holders and welcomes correspondence from those they have been unable to contact.

A Library of Congress record exists under LC control number: 80008022

ISBN 13: 978-1-138-94465-7 (hbk)
ISBN 13: 978-1-315-67175-8 (ebk)

TRENDS IN ENERGY USE IN INDUSTRIAL SOCIETIES

Research Paper R-19

TRENDS IN ENERGY USE IN INDUSTRIAL SOCIETIES

An Overview

Joy Dunkerley

RESOURCES FOR THE FUTURE / WASHINGTON, D.C.

ACKNOWLEDGMENT OF SUPPORT AND LEGAL NOTICE

Trends in Energy Use in Industrial Societies: An Overview was financed by the Electric Power Research Institute (EPRI) of Palo Alto, California under Research Project 864-1. The findings and conclusions, however, are those of the author and not of EPRI or any of its member organizations.

Library of Congress Catalog Card Number 80-8022

ISBN 0-8018-2487-7

Distributed by The Johns Hopkins University Press, Baltimore, Maryland 21218

Manufactured in the United States of America

Published September 1980. $8.00.

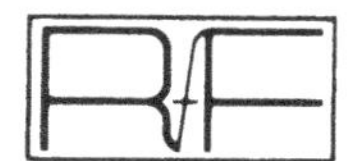

RESOURCES FOR THE FUTURE, INC.
1755 Massachusetts Avenue, N.W., Washington, D.C. 20036

Resources for the Future is a nonprofit organization for research and education in the development, conservation, and use of natural resources and the improvement of the quality of the environment. It was established in 1952 with the cooperation of the Ford Foundation. Grants for research are accepted from government and private sources only if they meet the conditions of a policy established by the Board of Directors of Resources for the Future. The policy states that RFF shall be solely responsible for the conduct of the research and free to make the research results available to the public. Part of the work of Resources for the Future is carried out by its resident staff; part is supported by grants to universities and other nonprofit organizations. Unless otherwise stated, interpretations and conclusions in RFF publications are those of the authors; the organization takes responsibility for the selection of significant subjects for study, the competence of the researchers, and their freedom of inquiry.

Research Papers are studies and conference reports published by Resources for the Future from the authors' typescripts. The accuracy of the material is the responsibility of the authors and the material is not given the usual editorial review by RFF. The Research Paper series is intended to provide inexpensive and prompt distribution of research that is likely to have a shorter shelf life or to reach a smaller audience than RFF books.

CONTENTS

LIST OF TABLES

FIGURES

FOREWORD

"Why is energy consumption per unit of output so much higher in the United States than in other advanced industrial countries with comparable living standards?" We at RFF became interested in this question shortly after the oil price rise of 1973/74 which directed attention at the prospects for conservation.

More than intellectual curiosity was involved. Policies were being proposed and challenged based on perceptions of the forces driving energy use per unit of output. Cross-country comparisons of energy use, we concluded, could enlighten the debate on such issues as the importance of developing new supply options, the response of energy use to higher energy prices, and the ability to alter energy use by command and control measures without affecting living standards. In more general terms, such comparisons could offer insights into the prospects for economic growth in an era of increasingly costly energy.

With the financial support of the Electric Power Research Institute (EPRI) RFF launched a detailed study of energy use in a number of countries for the year 1972. The results were published in How Industrial Societies Use Energy by Joel Darmstadter, Joy Dunkerley and Jack Alterman (Baltimore, Johns Hopkins University Press for Resources for the Future, 1977). This research answered some questions, posed others more carefully, and suggested further opportunities for research.

Numerous individuals and groups in different countries joined us in interest in this topic. We decided it would be useful to bring the experience of researchers from different countries to bear on the question of the uses and limitations of international energy comparisons in assessing conservation possibilities. To this end, RFF, again with EPRI financial support, organized a workshop in 1977. Its proceedings, under the editorship of Joy Dunkerley, were issued as an RFF Research Paper (R-10) entitled International Comparisons of Energy Consumption (1978).

The study underlying the present volume extends the previous work. Also sponsored by EPRI, it concludes, at least for the present, our work in this topic. We suspect, however, that other analysts will have more to say on the subject when additional data become available. Its purpose is to examine energy consumption over time in order to test contending explanations of energy trends and to throw additional light on the relationship between energy consumption and economic output.

RFF, in conducting these studies and publishing their results, has sought to contribute analytical rigor to a debate that has sometimes been carried out in an atmosphere of unrealism if not mysticism. I believe a notable measure of success has been achieved.

Milton Russell
Director,
Center for Energy Policy Research

August 1980

ACKNOWLEDGMENTS

This book is based on a report prepared for the Electric Power Research Institute (EPRI) in Palo Alto, California, entitled Trends in Energy Consumption in Industrial Societies, by Joy Dunkerley, Jack Alterman, and John J. Schanz, Jr.

The author of this book wishes to acknowledge with many thanks EPRI support of the research and development involved in this study and the encouragement given by the EPRI Project Manager, Albert N. Halter of Energy Analysis and Environment. She also wishes to acknowledge her great debt to the coauthors of the expanded version of the report, Jack Alterman and John J. Schanz, Jr. Jack Alterman was the author of a comprehensive study of U.S. energy-output ratios, whose conclusions are extensively reported here. John J. Schanz, Jr., was the Principal Investigator of the Project. The names of these two close colleagues would normally, except for their own most generous insistence, have appeared as coauthors of this volume.

Other colleagues from Resources for the Future contributed substantially. In particular, Joel Darmstadter provided valuable advice on the design and conduct of the study. Irving Hoch set up the regression programs, which were implemented by Judith Drake. Linda Sanford developed the extensive data base used in this study and also ran many of the regressions. Jo Hinkel was a most skilled and supportive editor. Most of the RFF Center for Energy Policy Research secretaries assisted in the typing of the manuscript. Special thanks for the final version are due to Mae Barnes, Lorraine Van Dine, Angela Blake, and Nhan Nguyen.

Useful comments and suggestions were received from Lee Schipper of the Lawrence Berkeley Laboratory, David O. Wood of the Massachusetts Institute of Technology Energy Laboratory and Sloan School of Management, and Leonard Gianessi of RFF.

TRENDS IN ENERGY USE IN INDUSTRIAL SOCIETIES

INTRODUCTION

This present study is a shortened version of a report _Trends in Energy Use in Industrial Countries_, prepared for the Electric Power Research Institute (Research Project 864-1, January 1980). The EPRI report contained, in addition to the text given here, nine technical and methodological Appendices including two detailed case studies of energy intensity in Sweden and the United States. The expanded version of the report is available from EPRI. This shortened version is designed to appeal to the reader who has an informed interest in this topic but who does not require the extent of methodological information contained in the longer report.

Trends in Energy Use in Industrial Societies grew out of a previous study (also financed by EPRI), _How Industrial Societies Use Energy_ (1). This study analyzed differences in the amount of energy consumed relative to output or income in nine industrial countries (the United States, Canada, France, West Germany, Italy, the Netherlands, the United Kingdom, Sweden, and Japan) for one year, 1972. In brief, that study concluded:

1. Energy consumption relative to total domestic output was pervasively higher in the United States and Canada than it was in other industrialized countries. The United States for example consumed 50 percent more energy relative to output than the average of the Western European countries and Japan, and 100 percent more than the lowest consumers (France and Japan).

2. This higher U.S. energy consumption relative to income is caused in part, but not entirely, by structural differences between the United States and the Western European countries and Japan. In this context, structural differences include factors such as the size of the industry sector, patterns of personal mobility, and the nature ofthe housing stock. These structural factors account for perhaps 40 percent of

higher U.S. consumption. Energy-intensity factors--the differing amounts of energy used in performing an activity, such as a passenger-mile, or producing unit of industrial output--accounted for the balance.

3. Underlying the higher U.S. and Canadian energy consumption relative to income are substantially lower energy prices. Variations in tastes, geography, and climate also play a part.

The present study examines how well these conclusions hold over time. With regard to our first conclusion--that the United States and Canada consume more energy relative to the GDP than other countries--we wish to find out whether this is a transitory or more permanent phenomenon and whether energy consumption relative to income among this group of countries was diverging or converging over this period.

Second, just as structural factors were found to account for part of the difference in energy consumption relative to output in 1972, we need to examine whether similar structural factors dominate changes in energy consumption relative to output over time, both for the aggregate economy and the constituent sectors.

A third purpose of this report is to assess more systematically the influence of varying energy prices on levels of energy consumption. In the previous study we concluded that higher European energy prices--ranging in 1972 up to 80 percent higher in Europe and Japan as compared with the United States and Canada--were of considerable significance in explaining the 15 to 50 percent lower energy consumption relative to income in those countries. Given the limited number of countries and data points in the previous report, we did not feel able to do more than suggest a strong causal influence between energy prices and energy consumption. The present study, which has a much more extensive data base, permits systematic investigation of the behavior of energy consumption in relation to energy prices, both in the aggregate and on the sectoral level. In addition, the role of institutional factors, particularly government policy, on energy consumption is assessed.

The present study therefore tests the conclusions of the previous study over time and more generally seek to throw additional light on the relationship between energy consumption and economic output.

Many of the conventions and definitions adopted in How Industrial Societies Use Energy are carried over into the present volume. Thus the same countries--the United States, Canada, France, Germany, Italy, the Netherlands, United Kingdom, Sweden and Japan--are included for the same reasons. Out of a preliminary list of twenty countries with relatively high levels of output per capita, nine were considered a sufficient number to permit generalized analysis, yet still small enough to allow the degree of detail felt essential.

Within the group of twenty relatively high-income countries, the nine selected were those countries with the largest economies (United States, Japan, West Germany, France, United Kingdom, and Italy); a small country (the Netherlands); a geographically large country (Canada); and a country with a large proportion of hydroelectricity (Sweden). These nine span a wide range of energy/GDP ratios. The highest, the United States and Canada, are over twice that of the lowest, France. The main group of industrialized countries not represented are the centrally planned economies.

In order to sharpen the analysis, we concentrate in this study more on comparing the average of the group of seven countries in Western

Europe and Japan (France, Germany, Italy, the Netherlands, UK, Sweden and Japan) with the United States. More detailed country analysis is contained in Appendix A of this study and in the Appendices contained in the EPRI version. We exclude Canada, like the United States a large energy consumer relative to GDP, in some summary tables in the interests of simplicity, but many of the conclusions applying to the United States could also apply to Canada. In the figures, only seven countries are included for visual effect. Canada is excluded because its experience is similar to that of the United States and Japan because of major differences between that country and the others.

The period covered by this report varies according to data availability. In some series it has been possible to go back as far as the mid-1950's. But for most of the analysis--and especially the more detailed analysis--the base period 1960 to 1976 is used. This period is usually broken into two sub-periods, 1960-73 and 1973-76 in order to trace the effects of the 1973/74 rise in oil prices on energy use.

As in How Industrial Societies Use Energy, our basic unit of measurement is the amount of energy consumed in relation to total output of Gross Domestic Product. This measure is more immediately informative in comparing differences in energy consumption over time and among countries than the more customary energy consumption per capita, which does not take into account differences in living standards.

Our energy consumption data are expressed in tons oil equivalent (toe), following the practice of the Organisation for Economic Co-operation and Development/International Energy Agency. Their series "Energy Balances of OECD Countries" explains methods of converting the individual fuels to the common oil equivalent measure. In the OECD/IEA system, unlike the United Nations, primary electricity is converted into tons oil equivalent as if it were generated from fossil fuels. This procedure raises the consumption of those countries, such as Canada and Sweden, with major hydroelectricity generating facilities. Our concern in this study is almost exclusively on the demand side of the energy equation. Issues of energy supply are mentioned only briefly, in Chapter 8, Energy Policy Responses.

Gross Domestic Product, which measures goods and services produced within the confines of national boundaries, rather than Gross National Product (GNP), which includes goods and services produced by nationals of a country on a world-wide basis, is used here. In practice the difference is not great, but GDP is a more appropriate measure to relate to our energy consumption data, which apply in general to energy consumed within national borders. Data on GDP available in national currencies have been converted to U.S. dollars using purchasing power parity rates of exchange (based on Irving B. Kravis, Zoltan Kennessey, Alan Heston and Robert Summers, <u>A System of International Comparisons of Gross Product and Purchasing Power</u> (Baltimore, published for the World Bank by the Johns Hopkins University Press, 1975)). This procedure yields a more realistic comparison of output levels among countries.

A more profound criticism concerns the use of GDP or GNP as a measure of output or welfare. As stated in <u>How Industrial Societies Use Energy</u>:

> The use of GDP in this study does not reflect confidence that growth in such a measure necessarily produces enhanced human welfare or perceived happiness, or that a shrinking national product need signify an erosion of such welfare. Gross domestic product and its constituent parts do, however, represent an objective reckoning of expressed market preferences, whereas most other standards for measuring the social product would almost certainly involve the imposition of intensely controversial and debatable individual value judgments. Such questions are exceedingly important elements in the contemporary debate over economic growth, environment, resource scarcity, and life-styles. We hope the results of this study--built on presently available, concrete, and quantifiable topics--can take their place with approaches emphasizing other standards of relevance. (1, p. 11).

This study is the third in a series published by Resources for the Future with the financial support of the Electric Power Research Institute. The first was _How Industrial Societies Use Energy: A Comparative Analysis_. The second was _International Comparisons of Energy Consumption, the Proceedings of a Workshop_ (ed. Joy Dunkerley, Research Paper R-10, Resources for the Future, Washington, D.C. 1978), covering a survey of existing studies, technical and methodological problems in international comparison, and the uses and limitations of international energy comparisons. Both RFF and EPRI hope that with these three volumes they have contributed some analytical balance to a hotly debated topic which has closely influenced views on energy conservation potential and strategy.

REFERENCES

1. J. Darmstadter, J. Dunkerley, and J. Alterman. _How Industrial Societies Use Energy: A Comparative Analysis_ (Baltimore, Md., Johns Hopkins University Press for Resources for the Future, 1977).

Chapter 1

TRENDS IN ENERGY OUTPUT RATIOS

Numerous studies testify to a close relationship between energy consumption and economic output. As economic output increases, so generally does energy consumption. But there is little evidence that energy consumption and economic growth move together in exactly the same proportions. On the contrary, recent experience in industrialized countries has shown that, both among countries and at different times within the same country, a given increase in economic output is associated with a wide range of change in energy consumption--sometimes a more than proportional increase, sometimes less, and sometimes a similar change.

A convenient way of expressing this relationship between energy consumption and economic growth is the energy/GDP or energy output ratio. This ratio, derived by dividing a country's energy consumption by its Gross Domestic Product (GDP) gives the amount of energy consumed relative to a given value of output. (In this study, the energy output ratio is expressed as energy consumed in tons of oil equivalent per million dollars of GDP).[1] The higher the ratio, the more energy-intensive the economy. If energy consumption and economic output both rose

[1]Primary electricity is aggregated to tons oil equivalent as if it were generated from fossil fuels and, therefore, it includes hypothetical heat losses. This procedure raises the apparent energy consumption of those countries such as Canada and Sweden which have a large primary electricity (hydro) component in their energy supply.

(continued)

and fell in exactly the same proportion in any one country, the energy/GDP ratio would be the same in all years. And if, among countries, differences in energy consumption were proportional to economic output, then all countries would have the same energy/GDP ratio. But as figure 1-1 indicates, this is clearly not the case. On the contrary, considerable diversity among countries is revealed. Of these countries, one (the United States), though displaying some variation within and between sub-periods, remains fairly steady over the period as a whole; three (the United Kingdom, West Germany, and France) experienced declining ratios (particularly sharp in the case of the United Kingdom and West Germany), and the remaining three (Sweden, Italy, and the Netherlands, but particularly the last two) experienced rising ratios. This diversity of trends in energy/output ratios among highly industrialized countries with broadly similar economic structures implies at first sight that there is considerable flexibility in energy requirements

---

(continued)

The measurement of total output (GDP), expressed in constant prices for each country, is converted to U.S. dollars by using estimated purchasing power parity rates of exchange, and is based on work by Kravis and coauthors (1). These rates of exchange seek to equate the values of all goods between countries rather than the values of internationally traded goods, which are those measured by market rates of exchange.

The general effect of using purchasing power parities rather than exchange rates has been to raise the level of other countries' GDP relative to that of the United States.

Figure 1-1. Energy consumption per unit of gross domestic product, 1953-76

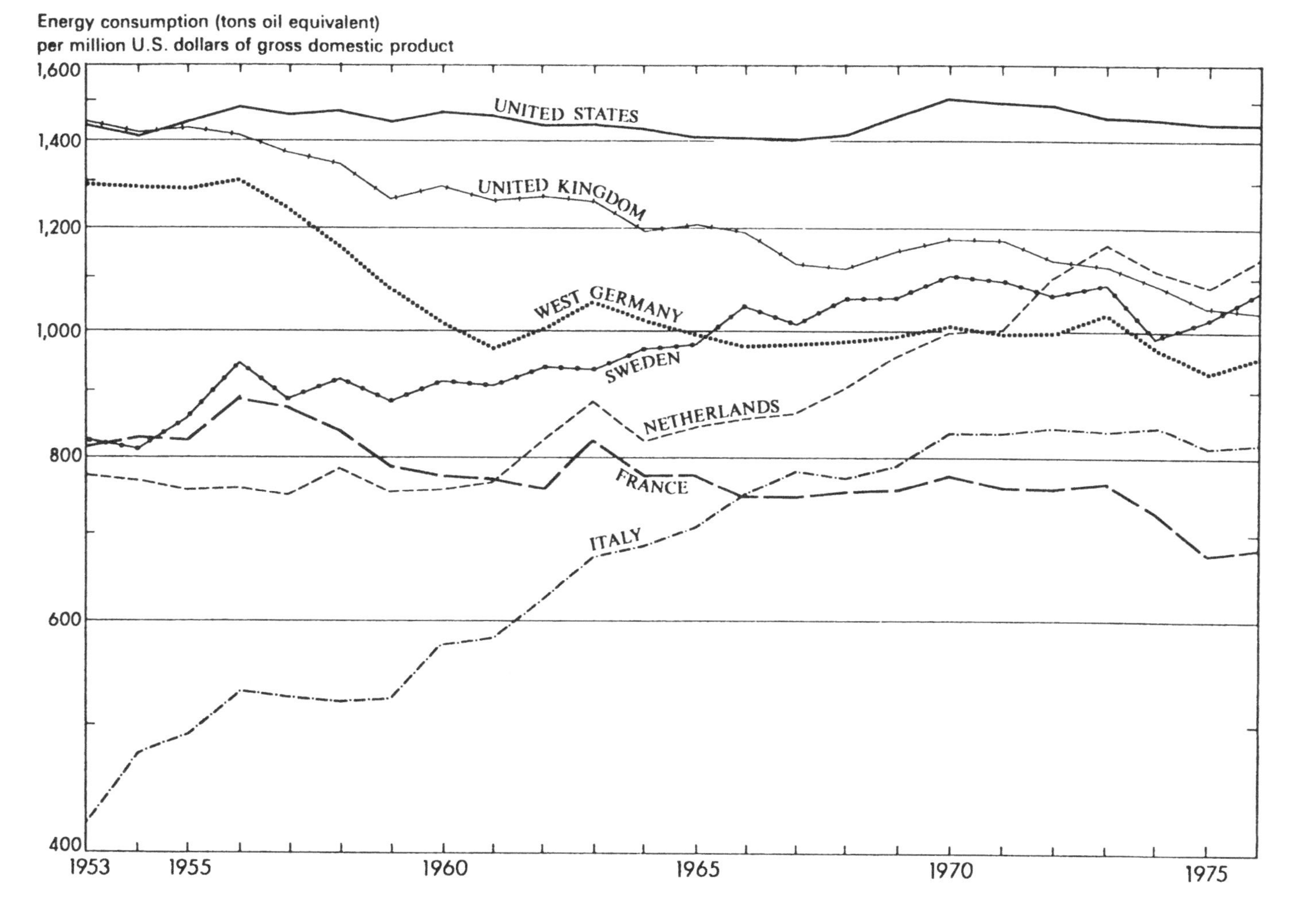

for a given increase in output, both among countries and over time. Far from a one-to-one relationship between energy consumption and output, there is a wide range of possibilities covering both less-than-proportional and greater-than-proportional energy inputs for a given change in output.

Figure 1-1 also indicates that the United States, among the countries depicted here, consumed over the decades the most energy relative to output. The comparatively high ratios of the United States, as compared with those of the other countries, seem to have been a permanent feature of the last twenty years.

On the other hand, considerable differences in ranking between the Western European countries took place. At the beginning of the period, the countries were grouped into two categories. These were, first, countries with high ratios including not only the United States and Canada, but also the United Kingdom and West Germany; and, second, a closely grouped set with lower ratios--Sweden, the Netherlands, and France. Italy was far below the others. Over the period, the ratios of the European countries with high energy/GDP ratios fell, and those of the European countries with low energy/GDP ratios rose. By the end of the period, this group of countries, which had widely different energy intensities in 1953, had drawn closely together, and several changes in ranking had taken place, The Netherlands now (1976) has the highest ratio and France the lowest.

Within this longer period, there were clearly three distinct subperiods: from 1953 to about 1965, from 1965 to 1973, and from 1973 to 1976. In the first period, very rapid changes in energy/output ratios in both

of European countries and between the European countries as a group and the United States. In the following years, while individual experience varied, energy/output ratios stabilized, and there were few immediately perceptible trends. This relative stability came to an end in 1974, when rising oil prices and worldwide recession led to declines in all countries' energy/output ratios.

These trends, based on aggregate energy consumption data, obscure changes in the end uses of energy over the period. Comparable data on the major end-use sectors for all countries unfortunately are not available over the longer period from 1953. Data are available, however, from 1960, permitting more detailed analysis for the last sixteen years. In relation to the longer-term trends, this more detailed data covers two of the subperiods--the 1960s up to 1973, when energy/output ratios were stabilizing, and the post-1973 period, when they fell.[2]

Five major end-use sectors are usually identified as follows (see Table 1-1):

1. Losses. This item includes heat losses incurred in the generation of thermal electricity (including imputed heat losses attributed to hydroelectricity; see footnote 1 in this chapter for an explanation), losses incurred in the production of manufactured, or town gas, and energy used by the energy industries such as the use of gas for pumping gas, electricity transmission and distribution losses, and consumption by oil refineries. Of this sector, heat losses predominate. Losses have tended to decline as a proportion of total consumption in many countries, with

---

[2]The two series of data are drawn from different sources. The series beginning in 1953 comes from the UN Statistical Office (2), and the series beginning in 1960 from the International Energy Agency/ Organisation for Economic Co-operation and Development (3). While there are difference in conventions and conversions between these two series, they are reconcilable and yield similar trends in aggregate energy consumption.

Table 1-1. Percentage of Energy Consumption by Sector

	United States	Canada[a]	France	Germany	Italy	The Nether-lands	United Kingdom	Sweden[a]	Japan
1960									
Industry	26.8	22.2	32.2	33.3	34.7	25.9	28.5	31.9	38.6
Residential-commercial	26.2	23.3	22.5	23.7	20.4	29.7	28.8	23.4	11.8
Transport	23.1	16.7	12.9	10.9	13.6	12.8	12.4	10.6	12.4
Nonenergy	3.7	3.1	2.6	2.2	4.0	1.0	2.6	3.0	1.9
Losses	20.1	34.8	29.8	29.9	27.2	30.6	27.7	31.1	35.3
1970									
Industry	25.5	22.4	33.2	28.7	33.2	24.1	26.4	30.4	37.9
Residential-commercial	26.1	26.0	24.4	28.4	21.7	31.5	25.0	30.3	16.6
Transport	22.2	17.7	14.1	11.8	14.3	12.7	12.6	11.2	11.3
Nonenergy	4.1	3.9	5.5	6.2	8.2	8.5	4.8	3.8	11.1
Losses	22.1	30.0	22.8	24.9	22.6	23.2	31.2	24.3	23.1
1976									
Industry	17.5	20.9	26.7	25.1	27.6	22.4	24.6	28.2	32.1
Residential-commercial	26.5	23.9	27.9	29.6	26.8	33.7	25.3	29.6	17.6
Transport	24.4	19.2	16.8	13.2	13.4	12.6	24.7	11.2	12.4
Nenenergy	5.4	3.4	5.5	6.0	10.4	13.8	5.1	4.7	8.1
Losses	26.2	32.5	23.2	26.0	21.8	17.4	30.3	26.3	29.7

Source: International Energy Agency/Organisation for Economic Co-operation and Development, Energy Balances of OECD Countries various issues (Paris, OECD, 1978).

[a]For Canada and Sweden, losses include heat losses imputed to the larger hydroelectric output, and are therefore not strictly comparable with the other countries.

the exception of the United States, where they have increased. Typically, they now account for about 25 percent of total consumption as compared with about 30 percent in 1960.

2. Industry, which for our industrialized countries accounted for about 30 percent of total energy consumption in 1960, declined to nearer 25 percent at the end of the period. The iron and steel industry is invariably the major single energy consumer within this group.

3. Nonenergy uses, consisting of petroleum feedstocks for the petrochemical industry and energy consumption incurred in the production of a wide range of nonenergy products such as lubricating oils, bitumen, and so on. Though accounting for only a small proportion of energy consumption, its share has grown sharply over the period (from an average of 3 percent to about 7 percent in 1976).

4. Transport, including energy consumption in passenger and freight transport by road, rail, air, and by inland navigation. Road transport dominates. This sector has, except in the United States, tended to account for 10 to 15 percent of the total. The U.S. share is significantly higher at 23 percent.

5. Residential-Commercial uses, which also include a wide variety of miscellaneous uses such as agriculture, handicrafts, government, street lighting. The major use in this category is space-conditioning which, on average, accounts for 60 to 70 percent of the total. The share of this sector has risen from about 23 percent of the total to about 27 percent.

The trends in sectoral energy consumption which underlie aggregate energy consumption are illustrated in table 1-2. Here, sectoral energy consumption data are given in relation to the GDP for the United States and for an (unweighted) average of seven other industrial countries (France, Germany, Italy, the Netherlands, the United Kingdom, Sweden, and Japan).[3] The first thing to notice about these trends in sectoral

---

[3]Appendix A gives more detailed tables of trends in energy consumption, distinguishing individual countries.

Table 1-2. Trends in Sectoral Energy Consumption Relative to the GDP

(in tons oil equivalent per million $ GDP)

	1960	mid-1960s[a]	1970	1973	1976
U.S. total	1,408	1,351	1,453	1,401	1,360
Losses	283	297	321	351	357
Industry	378	363	371	297	239
Nonenergy	52	43	59	79	73
Transport	325	305	323	326	332
Residential-commercial	370	342	379	348	360
Seven industrial countries' total[b]	917	942	991	1,001	954
Losses	277	260	246	243	231
Industry	293	284	299	280	252
Nonenergy	22	45	67	75	81
Transport	111	117	124	129	129
Residential-commercial	214	236	256	275	265
Seven industrial countries' total (U.S.=100)[b]	65	70	68	71	70
Losses	98	88	77	69	65
Industry	78	78	81	94	105
Nonenergy	42	105	114	95	111
Transport	34	38	38	40	39
Residential-commercial	58	69	68	79	74

[a]Average of 1964, 1965, and 1966.

[b]France, Germ ny, Italy, the Netherlands, United Kingdom, Sweden, and Japan. Canada is excluded to facilitate comparison between the U.S. on the one hand and the European countries and Japan on the other.

energy consumption is that until 1973 there is considerable divergence between them--some increased, some decreased, and some stayed relatively constant.

The trends in aggregate energy consumption described above might have been the result of similar trends throughout the main end-use consumption sectors, but this was clearly not the case. In each country some sectors expanded and others contracted. But within the divergent record, a common factor shared by both the United States and the other industrial countries was the decline in energy consumption by the industrial sector relative to the GDP. On the other hand, energy consumption for nonenergy uses (feedstock for petrochemicals, and so forth) rose.

A second feature of these trends is the difference between experience in the United States and the seven other industrial countries. In the United States, energy losses relative to the GDP rose over the period, and fell in the other countries. On the other hand, transport and residential-commercial use remained fairly stable in the United States, but increased steadily in the other countries.

On balance, there appears to be some tendency for energy consumption patterns in the industrial countries of Western Europe and Japan to approach the higher-intensity U.S. consumption patterns of the United States as the incomes of Western European countries approached U.S. levels. This is noticeable in both the industrial sector and the residential-commercial sector. The biggest difference now remains in the transportation sector, but even here there are signs that the European countries may be approaching U.S. comsumption patterns.

Since 1973, energy consumption, though continuing to rise slowly in absolute terms, has risen much less slowly than economic output with the result that energy/output ratios in all countries considered have fallen (see table 1-2). This marks a distinct break with the past. Prior to 1973, energy consumption relative to output had shown mixed results--rising in some countries and falling in others. After 1973, the ratio fell in all countries.[4]

The most striking way of presenting this change is to consider the elasticity of energy consumption with respect to the GDP. Thus International Energy Agency (IEA) data indicate that, up to 1973, for IEA member countries as a whole, the ratio of growth in energy consumption to growth in GDP was just over one. That is, a given increase in the GDP was associated with a similar rise in energy consumption. In the years 1972 and 1977 this ratio fell to 0.5--which means that a given increase in the GDP was associated with only one-half the increase in energy consumption.

The experience of individual countries varies considerably. Thus, on the basis of the same data, the U.S. elasticity falls from 1.11 in 1960-72, to 0.34 after 1972. Insofar as energy/GDP ratios can be taken as an indication of conservation achieved, the record of the United States is, in general, average. Some countries achieved more and some less. It is, however, perhaps early to make a final assessment of

---

[4]Post-1973 trends in energy/output ratios have been extensively treated elsewhere. Compare International Energy Agency/Organisation for Economic Co-operation and Development (4); Sawhill and coauthors (5); Erb (6); Rodekohr (7); and current work on energy conservation and its impact on the economy by the UN Economic Commission for Europe.

these trends, as the energy/GDP ratios varied considerably even after 1973, depending on the timing of recession and recovery. Thus, the ratios generally fell in 1974 and 1975 but rose again in 1976, though they still remained generally below 1973 levels. The main sectoral change contributing to this decline in aggregate energy consumption relative to output was a sharp decline in industrial energy use.

This chapter has had two themes. The first is the diversity among countries with regard to energy consumption. We have seen that even within a group of industrialized countries with similar levels of income, the amount of aggregate energy consumed in relation to a given value of output (the energy/output ratio) varies considerably--both among countries and within any one country over time. The findings on aggregate energy consumption are confirmed by an analysis of trends in sectoral energy consumption. At similar levels of output, countries consume very different amounts of energy in, for example, transport and residential-commercial uses.

The second theme is that despite this diversity there is evidence that energy consumption patterns are converging in some important respects. In particular, energy consumption patterns in European countries seem to be approaching those of the United States. This is particularly marked in the residential-commercial sector. Since 1973, energy/output ratios in all countries considered have fallen. This marks a distinct break with the past and can be attributed on a sectoral basis to a sharp decline in industrial energy use relative to the GDP.

In subsequent chapters, we will investigate the reasons for such differences in aggregate and sectoral energy consumption. In particular, we will examine the influence of changes in the structure of the fuel supply systems, in the structure of economic activity, in differences in energy prices, and in energy policies.

REFERENCES

1. I. B. Kravis, Z. Kennessey, A. Heston, and R. Summers, A System of International Comparisons of Purchasing Power (Baltimore, Md., Johns Hopkins University Press, 1975).

2. UN Statistical Office, World Energy Supplies, various issues. (New York, UN).

3. International Energy Agency/Organisation for Economic Co-operation and development, Energy Balances of OECD Countries, various issues (Paris, OECD).

4. International Energy Agency/Organisation for Economic Co-operation and Development, Energy Policies and Programs of IEA Countries, 1977 Review (Paris, IEA/OECD, 1978).

5. J. C. Sawhill, K. Oshima, and H. W. Maull, Energy: Managing the Transition, Trilateral Commission Triangle Papers 17 (Paris, Trilateral Commission, 1978).

6. R. D. Erb, "Will the Real Energy Policy Please Stand Up," The AEI Economist (August 1978).

7. M. Rodekohr, "Recent Energy Consumption Trends in the European Economic Community Countries," Energy Information Administration Technical Memorandum TM/IA/78-18 (Washington, D.C., U.S. Department of Energy).

Chapter 2

CHANGES IN ENERGY SUPPLY SYSTEMS

Changes in the form in which energy is supplied to the consumer can affect energy consumption significantly through differences in the efficiency with which energy is used. After all, what is important to most consumers is not so much the amount of energy, measured in thermal content, which is consumed, but the actual energy services or "useful" energy derived from the consumption. A move from lesser to greater efficiency in energy supply systems will, for a given amount of energy consumed, yield an increased amount of energy services. Conversely, a reduction in the energy efficiency of the supply system will require a larger energy input to satisfy the same demand for energy services.

Two major developments in the energy supply systems of the countries we studied took place between 1953 and 1976. The first was the widespread substitution of oil for coal in the Western European countries and Japan (1). In 1955 coal provided about 80 percent of European total energy consumption. By 1977, coal's share had fallen to about 24 percent, and oil's share had risen to 55 percent from 10 percent in 1950. Experience varied from country to country. While all started out heavily dependent on coal, some moved further and more rapidly to oil than others. Italy and Sweden, which had little domestic coal production, replaced coal imports by oil imports very quickly, and by the end of the period consumed only very small quantities of coal (7 percent and 4 percent of total energy consumption, respectively). Others, such as West Germany and the United Kingdom, which had large domestic coal industries, retained a much larger share of coal (about one-third) in total consumption (see table 2-1).

In the United States, where this transition had already run its course, there was relatively little change in composition of fuel supplies during these years.[1] Between 1953 and 1977, coal's share fell from just over 30 percent to 20 percent of the total, a much less precipitous fall than that experienced in Europe.

This contrast in coal-oil substitution between Western Europe and the United States contributed to the difference in energy/output ratios. In must uses, with the technology of the time, coal was a less efficient fuel than the others. As used in open fires for space heating, coal has an efficiency of between 10 and 20 percent, as compared with the 70 percent efficiency of fuel oil or gas boilers, and of over 90 percent efficiency for electricity , which has no flue losses.[2] In industrial uses, the differences between fuels are less marked, but coal still tends to have a somewhat lower efficiency than the others.[3]

---

[1]Coal's share of total energy consumption had fallen from 75 percent in 1915 to 29 percent in 1955.

[2]If the heat losses associated with electricity production are included, the efficiency of electricity would, of course, fall heavily. Thus the 90 percent output energy would require a total fuel input of 300 (including heat losses), making an overall efficiency of 90/300 x 100 = 30 percent. In fact the comparison is more complex if a heat pump is used.

[3]In modern industrial equipment, the efficiency of coal-fired boilers is as high as those for other fuels. But in the period we are dealing with here, the move from coal to oil would be also a move from old (and thermally inefficient) coal-using equipment to new (and efficient) oil-using equipment.

Table 2-1. Percentage of Coal in Total Energy Consumption

Years	United States	Canada	France	Germany	Italy	The Nether-lands	United Kingdom	Sweden	Japan
1953	32.8	39.4	70.9	92.2	32.2	79.6	89.1	28.8	65.5
1955	30.2	29.4	68.0	89.3	27.0	72.7	87.2	23.4	60.3
1960	23.8	16.1	57.0	76.1	17.0	55.2	76.8	11.9	51.7
1965	23.3	13.6	42.1	56.7	11.8	33.9	66.0	6.9	33.9
1970	20.3	10.7	26.4	39.9	8.7	11.3	50.7	4.4	24.5
1974	19.8	10.0	19.8	35.6	7.4	5.7	36.3	4.9	20.2
1977	22.0	12.2	20.0	31.7	6.9	5.6	37.6	4.3	17.8

Source: United Nations, World Energy Supplies, series J, various issues.

Finally, in rail transport, coal takes five times more thermal input to produce the same output (in terms of passenger or ton miles) as diesel fuel.

A move from coal to oil means, therefore, that for a given thermal input, more energy services or "useful" energy was being achieved. In other words, the conventional measurement of energy consumption in terms of heat content (as in figure 1-1) in those countries which were undergoing widespread conversion from coal- to oil-fired equipment would underestimate the energy services or "useful" energy derived.[4]

A second aspect of the changing structure of energy supply systems in these countries that is relevant to the interpretation of energy/output ratios is the size of the transformation or energy losses sector. It will be recalled from chapter 1 that this sector comprises energy used up in converting raw energy to a form suitable for use by the final consumer. It includes, for example, energy used in energy production (coal mining, oil and gas drilling), in energy conversion (refinery operations, electricity generation, coking), and electricity transmission losses. Energy used in this sector accounts typically for between 20 and 30 percent of total gross consumption. As tables

---

[4]This effect was first noted in the UK industrial sector by Adams and Miovic (2). They reported, "In line with conventional practice we have calculated energy elasticities with respect to industrial production and GNP....The elasticities based on unadjusted data are smaller than 1.0...After adjustment for fuel efficiency, they are equal to or greater than 1.0, suggesting that energy consumption increases proportionately or even more than proportionately with output".

1-1 and A-2 show, losses vary considerably amoung countries, with the United States, at least in the early part of the period, toward the lower end at 20 percent. Most of the other European countries at this time averaged about 30 percent. In the years that followed, however, losses as a percentage of total energy consumption rose in the United States but fell, often substantially, in Europe. By 1976, therefore, a convergence had taken place, with "losses" in both the United States and most Western European countries accounting for about one-quarter of total consumption. Table 2-2 illustrates the substantial increase in losses relative to the GDP in the United States in contrast to the net decline or more modest increase in the other countries. Thus, by 1976, U.S. losses were substantially higher than those of the other countries, whereas in 1960 they had been about the same level.

What caused this relatively sharp increase in U.S. losses? In order to answer this question, it is necessary to examine more closely the categories which make up total losses. Electricity losses (heat losses only, transmission and distribution losses are included in the own use category) account consistently for by far the largest part--two-thirds or more--of total losses. In the United States, electricity losses were the dominant force behind the rise in total losses. In Europe, the picture is more complex. Although frequently electricity losses determined overall losses, there were in some countries other decisive factors--a major example is the phasing out of town gas installations as natural gas came on-stream.

Table 2-2. Energy Consumption in Transformation Losses, 1960 and 1976 (tons oil equivalent/million $ GDP)

Year and Sector	United States	France	Germany	Italy	The Netherlands	United Kingdom	Japan
1960							
Total losses	283	245	303	172	252	337	306
Electric heat losses	191	168	164	132	128	203	221
Refinery losses	26	28	17	14	71	32	20
Own use and town gas losses	66	48	123	36	53	105	65
1976							
Total losses	357	165	253	186	207	306	210
Electric heat losses	257	108	187	136	132	217	142
Refinery losses	25	28	40	43	53	38	36
Own use and town gas losses	75	29	26	8	22	51	33

Source: International Energy Agency/Organisation for Economic Co-operation and Development, Energy Balances of OECD Countries, various issues (Paris, OECD).

Note: Canada and Sweden are not included because of their large hydroelectric component, which makes comparison with other countries difficult. The own-use category includes electricity transmission and distribution losses.

Those trends in electicity heat losses are, in turn, a product of two other factors, the amount of electricity generated, and the efficiency with which it is generated. All countries witnessed a rapid increase in electricity generation (see table 2-3), in all cases well in excess of the rise in GDP. In this respect, the United States differed little from the other countries with a compound rate of growth only a little higher than the average for the other countries.

But this trend that would, other things being equal, tend to increase electricity losses (relative to the GDP) in all countries appears to have been offset in many European countries by very sharp increases in generating efficiencies. While no doubt some margin of error exists in average efficiencies quoted here which are derived implicitly from fuel inputs and outputs of generating plants, there seems even so to be sufficient difference between the United States and other countries to suggest that there was, in fact, a much more rapid increase in generating efficiencies in these countries. This is a topic which merits further investigation.

These developments in fuel supply systems are of significance both to trends in energy consumption over time, and to the relative standing of the different countries with regard to their energy consumption. The generally growing efficiency of the European energy supply system over this period implies that consumption of energy services, or "useful" energy, was growing more rapidly than was suggested by data on gross energy inputs measured in heat content as used in figure 1-1. The same developments, combined with the generally stable energy efficiency of

Table 2-3 Electricity Consumption Relative to the GDP, and Implicit Generating Efficiencies, 1960 and 1976

Year and Sector	United States	France	Germany	Italy	The Nether-lands	United Kingdom	Japan
1960							
Electricity consumption relative to GDP (000 kWh/million $GDP)	1,244	683	850	713	626	995	1,100
Implicit generating efficiency (%)	36	26	30	32	30	30	29
1976							
Electricity consumption relative to GDP (000 kWh/million $GDP)	1,818	820	1,250	1,030	1,351	1,351	1,194
Implicit generating efficiency (%)	38	39	36	40	41	35	39

Source: International Energy Agency/Organisation for Economic Co-operation and Development, *Energy Balances of OECD Countries*, various issues (Paris, OECD).

the United States, imply that, measured in terms of useful energy, the European countries and Japan were drawing closer to the United States.

Given the potential importance of the changing efficiency of the energy supply system to the analysis of energy consumption in industrialized countries, we have converted gross energy consumption data measured on a heat content basis to "useful" energy. First, we deduct energy losses sustained in converting raw energy to a form in which it is delivered to final consumers, and, second, we apply coefficients of efficiency (see appendix B) to the various fuels used in major end uses.

This procedure yields higher "useful" energy for those countries where the fuel mix is concentrated on fuels that are highly efficient in use, such as gas and oil. As efficiencies for the same fuel differ among sectors, this procedure also incorporates changes in sectoral composition of energy consumption. Like changes in the fuel mix, these changes were much more marked in the countries of Europe than in the United States and Canada.

Even when sectoral variation is included, methods of adjustment for thermal efficiency rather than for thermal content of the different fuels are necessarily very crude. Fuels are not perfectly substitutable even on a thermal basis--for example, some require more expensive equipment in use. Furthermore, some fuels--and electricity is the prime example--have valuable qualities in addition to their thermal content. While it has not been possible to take these other qualities into account, we feel nonetheless that even imperfect estimates of "useful" energy, or energy services, give better insights into the underlying demand for energy than the more frequently used gross consumption data.

The effect of changes in energy-supply systems on energy consumption relative to the GDP is seen in table 2-4. For example, the gross data for France show a declining trend in the energy/output ratio over the period 1960-73. On the other hand, the adjusted data--that is, "useful" energy--indicate a distinct upward tendency. Similar patterns hold for West Germany and the United Kingdom. For Japan, a fluctuating ratio with no perceptible trend is converted into a sharply rising trend. For some countries, of course, adjustment of energy consumption data makes little change. The United States and Canada remain very similar whichever set of energy consumption data is used.

This implies that the "useful" energy/GDP ratios of the European countries and Japan were increasing quite sharply relative to the United States during this period. Table 2-5 gives data on various measures of energy consumption relative to that of the United States. It shows first, what had already been noted, that up to the mid-1960's the gross energy/output ratios of European countries and Japan (combined unweighted average), though still well below that of the United States, were rising relative to the United States. However, after the mid-1960s there was little change in relative position. Useful energy intensities for European countries rose much more sharply to the mid-1960s and continued rising thereafter, implying a much more marked convergence between the

Table 2-4. Gross and "Useful" Energy/Output Ratios

(toe/million $ GDP)

Country	1960	1966	1970	1973	1976
United States					
Gross	1,408	1,338	1,453	1,401	1,360
Useful	673	635	707	673	629
Canada					
Gross	1,728	1,604	1,677	1,742	1,632
Useful	628	690	749	771	743
France					
Gross	823	766	775	808	710
Useful	318	335	380	397	365
Germany					
Gross	1,011	982	1,015	1,026	973
Useful	379	436	490	520	465
Italy					
Gross	631	755	865	876	855
Useful	287	420	453	477	465
The Netherlands					
Gross	823	957	1,088	1,207	1,194
Useful	309	421	571	694	711
United Kingdom					
Gross	1,216	1,162	1,171	1,119	1,010
Useful	447	456	492	493	443
Sweden					
Gross	1,050	1,110	1,114	1,121	1,123
Useful	475	507	571	589	565
Japan					
Gross	867	869	908	848	800
Useful	349	440	499	482	398

Source: Gross energy consumption is total unadjusted energy consumption. Figures are from International Energy Agency/Organisation for Economic Co-operation and Development, Energy Balances of OECD Countries, various issues (Paris, OECD). Useful energy consumption excludes energy losses and is adjusted to take into account the differing thermal efficiency of fuels using coefficients from appendix B. Both series of energy consumption data are divided by GDP, taken from Organisation for Economic Co-operation and Development, National Accounts of OECD Countries, various issues (Paris, OEDC), converted to dollars using purchasing parity rates of exchange.

Table 2-5. Energy Output Ratios at Different Definitions of Energy Consumption (U.S. = 100)

Countries	1960	1966	1970	1973	1976
Europe and Japan (unweighted average)					
Gross	65	70	68	71	70
Useful	54	68	70	78	77

Source: Table 2-4.

two patterns of consumption. Since 1973, though the trend in energy intensities has declined, there has been little change in the relative rankings of countries.

Sectoral energy intensities calculated in terms of "useful" energy confirm and intensify the trends described in chapter 1. Consumption in residential-commercial uses relative to the GDP, which in gross terms had risen gradually, rose much more rapidly when expressed in useful terms. This is because in Europe coal was being replaced by more efficient gas and oil in the residential-commercial sector. Correction for thermal efficiency of fuels also further increases energy consumption in transportation. In the industrial sector, however, energy consumption continues to decline in most countries relative to the GDP, despite the correction for thermal efficiencies.

This chapter has examined the effect of changing energy-supply systems on energy consumption. It was found that the energy-supply systems of the European countries were moving in an energy-efficient direction because of declining conversion losses and a continued move away from coal to oil. This means that the underlying demand for energy, as indicated by the demand for energy services, was rising faster than suggested by data on gross energy consumption. As, during the same period, the U.S. system remained reasonably stable, the European pattern of energy consumption appeared to be moving toward that of the United States, both in quantities consumed and in sectoral distribution. Part of the differences in levels of energy consumption relative to the GDP among and within countries during this period can be attributed to differences and changes in the structure of the fuel supply sustems.

REFERENCES

1. Joy Dunkerley, The Switch to Oil and the Future for Coal in Western Europe, Resources for the Future Reprint No. 167 (Washington, D.C., Resources for the Future, 1978).

2. F. G. Adams and P. Miovic, "On Relative Fuel Efficiency and the Output Elasticity of Energy Consumption in Western Europe," Journal of Industrial Economics vol. 23, no. 1 (November 1965) pp. 553-554.

Chapter 3

EFFECTS OF CHANGING STRUCTURE OF ECONOMIC ACTIVITY

Differences in the composition of economic activity either among countries or within any one country over time can also affect the amount of energy consumed. To take a simple, if extreme, example: if one country, "Country A," has a large industrial sector, concentrating on steel-making, and a small agricultural output, its energy consumption relative to income or output will be higher than that of "Country B" with a large agricultural sector and small steel-making capacity. Even if "Country A" uses less energy than "Country B" per unit of output in both steel-making and agriculture, the predominance of an inherently energy-intensive activity will ensure a higher overall energy intensity.

In a previous study, How Industrial Societies Use Energy (1), we tried to take into account similar factors in explaining differences in energy consumption relative to output between countries. In that study, we distinguished between structural and intensity factors. To take the previous example, structural factors would include the size of the industrial sector and, within the industrial sector, the concentration on energy-intensive industry. Intensity factors would include the amount of energy needed to produce a given quantity of like products such as a ton of steel or a bushel of wheat.

If our analysis were limited to the composition and size of industrial output, however, it would fall short of covering the whole of energy consumption, since manufacturing plus agricultural energy consumption, for example, accounts typically for only between 20 and 30 percent of total energy consumption in industrialized countries. We attempted therefore to extend the structural-intensity analysis to the other sectors of energy consumption--that is, the residential-commercial and transport sectors. In these last two cases, the distinctions between structure and intensity are less clear-cut. For the transport sector, we classified as structure the number of passenger- and freight-miles traveled and the mode in which they were traveled. Intensity factors were limited to the amount of energy used in performing a passenger- or freight-mile. In the residential-commercial sector, we classified as structure the size and nature of the housing stock and differences in climate. Intensity was defined by the amount of energy used in heating or cooling a given house.

Following these definitions, we concluded that roughly 40 percent of the higher U.S. energy consumption relative to output was caused by structural factors, indicating that composition of output and analogous differences among countries explain a substantial part of the differences in their energy consumption levels relative to income.

These findings, based on a comparison among countries, suggested that similar compositional changes in the structure of economic activity might also affect trends in energy consumption relative to income over time. The large and complex data requirements of such an exercise limited most of this analysis to two countries--the United States and Sweden--in the ten-year period from the mid-1960s to the mid 1970s. (The detailed case studies for the United States and Sweden are available in the EPRI version of the study).

Here we concentrate on comparing the conclusions of the two case studies and provide additional analysis of the other countries where data are available.

As a point of departure, we compare briefly the differing trends in energy consumption over recent years in both the United States and Sweden. For purposes of comparison, we take energy consumption delivered to final consumers and at this stage exclude heat losses incurred in the generation of electricity for technical reasons. Because most U.S. capacity is thermally generated and most Swedish capacity is hydroelectric, it is difficult to find a satisfactory convention to permit comparison on the level of total primary energy consumption. If we look at energy consumption in the final consuming sectors, however, we observe considerable differences between the two countries (see table 3-1).

Swedish consumption relative to income was consistently lower than that of the United States. It was, however, rising steadily toward U.S. levels throughout the period. In 1960 Sweden used only 64 percent of the amount of energy consumed in the United States (both relative to income). By 1976 this had risen to 83 percent.

This convergence between U.S. and Swedish energy consumption levels was experienced in all sectors. Indeed, by the end of the period Sweden was consuming more energy relative to the GDP in both industrial and non-energy uses. The results for the industrial sector in 1976, however, may be disproportionately affected by cyclical factors.

In residential-commercial uses, Swedish consumption was also close to U.S. levels (relative to the GDP). The Swedish heating load is, of course, heavier than that of the United States because of the long, cold

Table 3-1. Energy/Output Ratios: United States and Sweden

(toe/million $ GDP)

Country	1960	1970	1973	1976
United States				
Total energy consumption (delivered to final consumer sector)	1,125	1,132	1,049	1,003
Residential-commercial	369	379	348	360
Transport	325	323	325	332
Industry	378	370	297	239
Nonenergy	52	59	79	73
Sweden				
Total energy consumption (delivered to final consumer sector)	725	844	864	872
Residential-commercial	246	338	336	332
Transport	111	125	124	126
Industry	336	338	349	316
Nonenergy	32	43	55	98
Sweden (U.S.=100)				
Total energy consumption (delivered to final consumer sector)	64	75	82	87
Residential-commercial	67	89	97	108
Transport	34	39	38	38
Industry	89	91	118	132
Nonenergy	62	73	70	134

Source: Based on International Energy Agency/Organisation for Economic Co-operation and Development, *Energy Balances of OECD Countries*, various issues (Paris, OECD); and Organisation for Economic Co-operation and Development, *National Accounts of OECD Countries*, various issues (Paris, OECD).

Note: The GDP data on which this table has been based are converted to dollars at purchasing power parity rates of exchange.

Swedish winters. Sweden, for example, has twice as many heating degree-days as the U.S. average.[1] The greatest difference between the two countries in energy consumption relative to GDP is in the transport sector, where Sweden, despite some increase over the years, is still well under (less than 40 percent) U.S. levels.

The case studies enabled us to enlarge upon these general trends in two ways: (1) by permitting further disaggregation of end-use sectors; and (2) by partitioning changes in energy consumption relative to income into a structural and an intensity component (see tables 3-2 and 3-3). Because the data requirements of this exercise are so taxing, it was possible to analyze both countries in detail for only two periods--the mid-1960s to 1972-73 and from 1972-73 to 1975-76.

With regard to additional detail concerning sectors, it was possible to disaggregate further the residential-commercial and the transport sectors for both countries. For both countries, in the period encompassing the mid-1960s to 1972-73, residential consumption was declining (relative to the GDP) while commercial consumption was increasing. After 1972-73, consumption in the commercial sector continued to increase, if slowly, in the United States, while it declined in Sweden, though by much less than the residential sector did. To summarize, within the larger sector, trends in the two constituent parts diverged, with the share of commercial consumption expanding rapidly.

---

[1]Degree-days measure the difference between actual temperature and the temperature at which heating is needed (the threshold temperature) in a 24-hour period.

Table 3-2. Analysis of Changes in U.S. and Swedish Energy Output Ratios, 1965/66 to 1972/73 (weighted annual average rates of change)[a]

(percent)

	United States 1966-73			Sweden 1965-72		
Sector	Total	Structure	Intensity	Total	Structure	Intensity
Residential-commercial	-0.07	0.06	-0.12	0.30	-0.10	0.39
Residential	-0.15	0.06	-0.20	-0.11	-0.19	0.08
Commercial	0.08	0.00	0.08	0.41	0.09	0.31
Transport	0.37	0.24	0.12	0.04	0.00	0.03
Passenger	0.35	0.23	0.11	0.07	0.03	0.04
Freight[b]	0.02	0.01	0.01	-0.03	-0.03	-0.01
Manufacturing[c]	-0.14	-0.31	0.17	0.12	0.29	-0.17
Thermal losses	0.65	0.65	0.00	0.35	0.44	-0.07
Energy industry	0.00	0.08	-0.06	-0.12	-0.12	0.00
Total of above	0.81	0.72	0.11	0.69	0.51	0.18
Other	-0.06	--	--	0.46[d]	--	--
Total	0.74	--	--	1.15	--	--

Source: Based on Joy Dunkerley, Jack Alterman, and John J. Schanz, Jr., Trends in Energy Use in Industrial Societies, EPRI Research Project 864-1 (Palo Alto, CA., Electric Power Research Institute, 1980) appendices E and F.

[a]For distinction between structural and intensity factors in Sweden and the United States, see the beginning of this chapter.

[b]For U.S. intercity freight transport.

[c]Excludes refinery operations, which are included in the sector entitled "Energy industry."

[d]Nonenergy uses.

Table 3-3. Analysis of Changes in U.S. and Swedish Energy Output Ratios, 1972/73 to 1975/76 (weighted annual average rates of change)[a]

(percent)

Sector	United States 1973-76			Sweden 1972-75		
	Total	Structure	Intensity	Total	Structure	Intensity
Residential-commercial	0.06	0.27	-0.21	-1.12	-0.75	-0.37
Residential	0.00	0.15	-0.15	-0.89	-0.60	-0.30
Commercial	0.06	0.12	-0.06	-0.22	-0.15	-0.07
Transport	-0.06	0.11	-0.18	-0.03	-0.07	0.04
Passenger	-0.13	0.15	-0.29	0.02	0.00	0.02
Freight[b]	0.07	-0.04	0.11	-0.05	-0.07	0.02
Manufacturing industry[c]	-0.91	-0.30	-0.61	0.05	0.11	-0.07
Thermal losses	0.24	0.27	-0.03	-0.64	-0.62	-0.02
Energy industry	-0.16	-0.06	-0.10	-0.27	-0.27	0.00
Total of above	-0.83	0.29	-1.13	-2.01	-1.60	-0.42
Other	-0.41	--	--	-0.42[d]	--	--
Total	-1.29	--	--	-2.40	--	--

Source: Based on Joy Dunkerley, Jack Alterman, and John J. Schanz, Jr., Trends in Energy Use in Industrial Societies, EPRI Research Project 864-1 (Palo Alto, CA., Electric Power Research Institute, 1980) appendices E and F.

[a] Distinction between structural and intensity differs in Sweden and the United States, see the beginning of this chapter.

[b] For U.S. intercity freight transport.

[c] Excludes refinery operations, which are included in the sector entitled "Energy industry."

[d] Nonenergy uses.

Within the transport sector, trends in the two major components, passenger and freight transport, also diverged. In the earlier period, the passenger transport segment rose much more sharply than that for freight; and, indeed, in Sweden it continued to rise even after 1973. In the United States, it declined in the 1973-76 period, both in absolute terms and relative to intercity freight.

The case studies also permitted the disaggregation of overall energy/GDP ratios into structural and intensity effects. Summary results for the United States and Sweden on a comparable basis (weighted annual average rates of change) for the two periods common to both (the mid-1960s to 1972-73 and from 1972-73 to 1975-76) are given in tables 3-2 and 3-3. These results must be interpreted with caution.

Sectoral definitions differ between the two countries, as do definitions of structure and intensity.[2] Furthermore, aggregate and sectoral energy consumption data derived from national sources are, in certain cases, difficult to reconcile with other series in this study--a constant problem when it is necessary to use more than one source of statistical data.

With these reservations, a number of conclusions can nonetheless be drawn. The first is that the patterns of change in the constituent parts

---

[2]The primary example of this difference occurs in the residential sector, where in the United States the structural or mix effect is defined by the changing proportion of consumer expenditures in total expenditures and the intensity effect by the amount of energy consumed per unit of consumer expenditure. In Sweden, on the other hand, structure is defined by changes in the stock of housing and household appliances, whereas intensity is defined as differences in energy that is used in connection with a given stock of houses and appliances.

of the energy/GDP ratio are very complex and do not lead to easy generalization. In the United States particularly, the small changes in aggregate energy consumption relative to output are the product of a large number of very small, frequently offsetting movements. For the United States, there is clearly no strong and simple explanation for the change in the aggregate energy/GDP ratio. For Sweden, the changes in the aggregate ratios were greater but, here again, they were the product of a large number of offsetting small changes.

Second, both structural and intensity factors contributed, generally in the same direction, to changes in the overall energy/GDP ratio. It should be borne in mind that this conclusion is based on the experience of only two countries over a limited time period. It does not apply, for example, to the U.S. experience in 1973-76.

This finding--that both structural and intensity factors contributed to changes in overall energy consumption--is confirmed by our earlier study (1), which found that the higher U.S. energy consumption as compared with the other industrial countries was caused both by structural and intensity factors.

This conclusion is also confirmed in a detailed study of the energy intensity of the Italian economy (2), which partitioned the total increase in energy consumption that took place in Italy between 1953 and 1972 into activity and intensity effects. Of the total increase in energy consumption, half is attributed to activity and half to increases in energy intensity.

While it can be concluded that both structural and intensity characteristics have contributed to changes in aggregate energy output ratios, it is not possible to generalize about which of these two factors has proved the stronger or which has acted systematically to increase or decrease the overall ratio. In Sweden, for example, structural characteristics dominated both the rise and fall in the aggregate ratio. In the United States, on the other hand, while structural elements were always positive, leading to an increase in the aggregate ratio, they did not always dominate. In Italy, intensity appears to have been dominant. And in the intercountry study (1), structure and intensity were found to contribute nearly equal shares.

Despite this mixed experience, some common characteristics of the experience in the United States and Sweden are revealed. We have already referred to the declining share of the residential segment within the residential-commercial sector and the rising share of passenger transport in total transportation. Another characteristic is the major contribution of electricity conversion losses to changes in the aggregate ratio. This stems (see tables 3-2 and 3-3) from the size of thermal generation relative to the GDP (a structural effect) rather than from any marked changes in heat rates (the intensity effect). Nonenergy uses, where distinguished, also contributed heavily to changes in energy/GDP ratios.

A further common feature was the decline in manufacturing energy intensities since 1972-73. In the case of Sweden, this was a continuation of the decline that had taken place in the earlier period. In the United States, on the other hand, industrial energy intensities rose between 1966 and 1973, though over the 1954-76 period as a whole industrial energy intensities declined.[3]

This decline in industrial energy intensities appears to be fairly general among industrialized countries, except in Italy where the rapid rate of growth in the industrial sector during this period included development of particularly energy-intensive industries such as petroleum refining and petrochemicals. Table 3-4 gives an index of energy consumption related to industrial production, showing a downward movement for most of the nine countries between 1960 and 1974.[4]

This index, which relates total industrial energy consumption to total industrial output, does not take into account the changing mix of industries within the industrial sector. It is therefore not possible to assess whether this decline is caused by intra-industrial shifts or by a decline in energy consumption per unit of like product. But some indications can be derived from the experience of the iron and steel industry, which typically takes about 20 percent of total industrial energy consumption. As a highly energy-intensive industry, it has a strong influence on total industrial energy consumption.

In the iron and steel industry, whose relative size varies from country to country, there has been a reduction in the energy intensity of the industry. That is, over time the amount of energy required to

---

[3]This subject has also been extensively analyzed elsewhere (3, 4, 5).

[4]Note that the use of value added by industry (in real terms) and industrial production as measures of output in this sector gives different and sometimes divergent results. Part of this difference probably lies in the way industrial activity data are calculated.

Table 3-4. Index of Industrial Energy Consumption Relative to Industrial Production

(1960=100)

Country	1966	1970	1974
United States	73	91	85
Canada	87	94	90
France	101	103	89
Germany	89	87	85
Italy	97	119	122
The Netherlands	103	100	95
United Kingdom	96	93	86
Sweden	88	91	78
Japan	75	77	71

Source: Data on industrial energy consumption are taken from International Energy Agency/Organisation for Economic Co-operation and Development, *Energy Balances of OECD Countries* (Paris, OECD, 1979), and adjusted to take into account thermal efficiency by coefficients in Appendix B. Industrial production data are from Organisation for Economic Co-operation and Development, *Economic Indicators* (Paris, OECD, 1979).

produce a ton of steel has fallen (table 3-5). If iron and steel is a good indicator of industry trends as a whole, then the fall in energy consumption per unit of industrial output would be caused primarily by a reduction in energy intensity per unit of product output. It is, however, risky to apply generalizations based on the experience of an old industry, such as the iron and steel industry, to the wider industrial experience.

Finally, the post-1972-73 analysis gives some indication of the effect that the 1973-74 rise in oil prices had on energy consumption. In both cases, the aggregate ratio declined, and the decline was fairly widely distributed among sectors. Energy intensities, in particular, showed consistent declines in both countries.

Conclusion

The lack of precision, inadequacies of data, and the necessity of making a large number of somewhat arbitrary decisions in defining what constitutes change in structure versus intensity preclude drawing strong conclusions from the case studies, but some tentative findings can be made.

The previous comparative analysis of industrial societies for the year 1972 showed that both structural and intensity characteristics contributed to variations in energy/output ratios between countries, and the results of the two case studies reported here confirm this finding. But it is not possible on the basis of these and other studies to generalize about which of these factors proved dominant, or which has acted systematically to increase or decrease the overall ratio. Common

Table 3-5. Energy Consumption Per Ton of Crude Steel Production (kilo/ton)

Country	1960	1966	1970	1974
United States	758	524	453	443
Canada	474	377	403	373
France	590	567	575	545
Germany	557	496	501	510
Italy	404	431	406	407
The Netherlands	626	574	494	491
United Kingdom	655	521	577	529
Sweden	519	483	431	398
Japan	468	450	454	435

Source: Data for energy consumption in the iron and steel industry are taken from International Energy Agency/Organisation for Economic Co-operation and Development, Energy Balances of OECD Countries (Paris, OECD, 1979). Data for crude steel production are taken from United Nations, Statistical Yearbook, various issues (New York, UN).

features of experience in the case studies, which can probably be generalized to the other industrial countries, are the sharp rises in energy consumption for the commercial sector, passenger transport, and nonenergy uses. Most countries also shared in the decline in industrial energy intensities.

Finally, the tendency for some ratios to alter their behavior after 1973 suggests that price and policy have had some impact. Therefore, in the next chapters we will turn our attention to the effect of differences in energy prices on energy consumption.

REFERENCES

1. J. Darmstadter, J. Dunkerley, and J. Alterman, How Industrial Societies Use Energy: A Comparative Analysis (Baltimore, Md., Johns Hopkins University Press for Resources for the Future, 1977).

2. O. Bernardini, "Structure and Technology as Determinants of Energy Demand in Post World War II Italy," in International Energy Agency/ Organisation for Economic Co-operation and Development, Workshops on Energy Supply and Demand (Paris, IEA/OECD, 1978).

3. The Conference Board, in cooperation with the National Science Foundation, Energy Consumption in Manufacturing (Cambridge, Mass., Ballinger, 1974).

4. G. C. Myers and L. Nakamura, Saving Energy in Manufacturing (Cambridge, Mass., Ballinger, 1978).

5. Stanford Research Institute International for General Electric Company, Industrial Energy Consumption Comparison: West Germany and the United States, April 1978 (Menlo Park, California SRI, 1978).

Chapter 4

THE EFFECT OF PRICE AND OTHER FACTORS ON ENERGY CONSUMPTION

The influence of energy prices on energy consumption is a question of major practical importance. Much current discussion of energy policy hinges on whether the rise in energy prices which has taken place in recent years is sufficient in itself to dampen consumption or whether additional measures are needed. In chapters 4, 5 and 6 we will therefore examine the influence of prices and other related factors on energy consumption.

Before proceeding to the analysis of the effect of energy prices on energy consumption, an important clarification must be made. So far in this study we have been taking as our basic measure of energy consumption the energy intensiveness of a country, that is, the amount of energy consumed in relation to total output or Gross Domestic Product (GDP). We have used this measure because it provides a shorthand description of energy consumption that has been corrected for differing levels of income or output. The other customary measure, per capita consumption, does not take into account variations in output levels and is not as immediately informative.

For the purposes of assessing the relative importance of different factors on energy consumption, however, the preferred measure of energy consumption is per capita consumption rather than the energy/GDP ratio. This is so because we cannot properly isolate the influence of prices on consumption levels without, at the same time, isolating the influence of income. By taking energy per unit of GDP as our measure of energy consumption, we are implicitly assigning a constant (unitary) proportional relationship between energy consumption and income. If in practice this does not hold, then the influence of prices will be distorted as they will be carrying some of the income effect. For this part of our analysis, therefore, we shall seek to explain changes in energy consumption per capita over time and between countries. Having done that, we shall recast our conclusions in terms of energy/GDP ratios in order to answer the question: What causes energy/GDP ratios to change over time and to differ between countries?

Note that we are referring here to useful energy consumption, that is, one of the factors--changes in the energy supply system--which would be expected to affect levels and changes in energy consumption, has already been incorporated into the measure to be explained. The effect of this adjustment, explained in chapter 2, is to increase energy consumption in the countries of Western Europe and in Japan between 1960 and 1973 more rapidly than if the adjustment had not been made (see table 4-1). In other words, the underlying demand for energy in these countries was probably increasing more rapidly than is indicated by gross energy data, but was to some extent masked by changes in the fuel supply. From 1973 to 1976 per capita consumption decreased in most countries, regardless of whether useful or gross energy is considered.

The next step is to inquire why this increase in "useful" energy consumption took place, and why it differed so much between countries, with particular reference to the effect of differing levels of incomes and energy prices.

Table 4-1. Annual Average Percentage Changes in Per Capita Energy Consumption and Gross Domestic Product for Selected Countries, 1960-76

	United States	France	Germany	Italy	The Netherlands	United Kingdom	Japan
<u>1960-73</u>							
Gross energy consumption	3.1	4.5	3.9	7.1	7.0	1.6	9.0
Useful energy consumption	2.9	6.1	5.5	8.3	9.1	2.8	11.2
GDP	3.1	4.7	3.7	4.4	3.9	2.3	9.2
<u>1973-76</u>							
Gross energy consumption	-0.9	-1.4	-0.5	-2.7	0.9	-2.5	0
Useful energy consumption	-1.5	-0.5	-2.0	1.9	1.9	-3.0	-2.7
GDP	0.3	2.2	1.5	1.3	1.5	0	1.2

Source: International Energy Agency/Organisation for Economic Co-operation and Development, Energy Balances of OECD Countries, various issues (Paris, OECD); and Organisation for Economic Co-operation and Development, National Accounts of OECD Countries, various issues (Paris, OECD).

Changes in Income and Output

Energy consumption is related to levels of income and output in the following ways. There is first a straightforward link between increases in the GDP and energy consumption as both cause and effect. An increase in output will require additional energy inputs, and the resulting increase in wages and salaries will, in turn, lead to increased energy use either through direct expenditure by households for energy such as gasoline, or through the energy content of additional goods and services consumed.

A second connection between output and energy consumption lies in the rate of growth of the GDP. Energy is used in conjunction with machinery and equipment (capital stock) in industry, in households, and in transport facilities. Energy consumption is therefore closely related to the size of this capital stock and its energy-using characteristics (such as production and transport technologies, infrastructure and social capital, and vintage of capital stock). The faster the rate of growth in the GDP the greater is the possibility of turning over energy-using capital stock. It might be thought that, given the decline in real prices of energy which took place up to 1973, capital stock of recent vintage might be more energy intensive than the old. Indeed, so far as private consumption is concerned, it appears that energy-using equipment--cars and household appliances--tended over time to become more energy intensive in use. Similarly, the generally lower U.S. energy prices resulted in a stock of consumer durables which are more energy intensive than those used in other countries.

On the other hand, industrial energy-using equipment of recent vintage appears, in some industries at least, to be more energy efficient than the old. This means that, in the industrial sector, a country with a faster rate of growth might have a newer and more energy-efficient capital stock than a country whose slower growth implies a higher proportion of older, less energy-efficient equipment. While comprehensive data on this important topic are not available, the recent vintage of capital stock was an important contributing factor in explaining the low energy consumption per unit of output in iron and steel and in cement in Japan, Italy, and West Germany--countries which underwent rapid growth since the 1950s (1, 2). And conversely, the high energy consumption in the low-growing UK and U.S. iron and steel industries appears to have been associated with the prevalence of older technology and equipment.

The reason behind the development of energy-saving processes and technologies in these industries, despite falling energy prices, may be that both the iron and steel and cement industries are highly energy intensive, and energy represents an untypically high proportion of total costs. High energy bills could therefore offer an incentive to consumers for adopting energy-saving technology as part of a general cost-minimization strategy.

Finally, as we saw in chapter 3, the composition of the GDP can affect energy consumption. A country with a large industrial sector, particularly if concentrated in heavy industries such as iron and steel, will consume more energy than a country with a small industrial sector that concentrates on lighter industries.

The close connection between energy consumption and output is generally confirmed by the experience of the past years (see table 4-1). Thus, the most rapid growth in energy consumption occurred in Japan, which experienced the most rapid rate of economic growth. And the slowest rate of energy growth took place in those countries (United States and United Kingdom) with the slowest rate of economic growth. The effect of rapid economic growth on energy consumption in Japan may have been reinforced by the disproportionate growth of its industrial sectors. The increase in energy consumption in this country might, however, have been even more rapid if the high rates of growth had not permitted the introduction of energy-saving technology.

Energy Prices and Energy Consumption

Changes in energy prices, both over time and among countries, will also affect energy consumption, influencing both the size and composition of the stock of energy-using equipment, and the energy-using characteristics of that stock. And, in the shorter run, energy prices can affect the rate of utilization of energy-using equipment.

Table 4-2 gives energy prices for all countries over the last twenty years. The prices are expressed in index form (U.S. 1972 = 100) and have been weighted in their sectoral detail by the quantities of the different fuels consumed within the country studied. They are, in other words, the unit cost index of a given amount of thermal content. This index applies to prices, including taxes, paid by consumers.

As the price in national currencies has been converted to a common unit, using purchasing power ratios rather than exchange rates, the individual country indexes may be, and are intended to be, compared with each other. The final index given here is obtained by calculating a weighted average of all three sectoral price indexes (industrial, residential, and transport) which were previously deflated by indexes of wholesale and retail prices, respectively. Therefore, the final index measures movements in real energy prices relative to other prices.

In interpreting this price data, a note of caution should be given. Energy prices are difficult things to measure. Gas and electricity are typically sold under a system of declining block rates in most countries so that it is difficult to know which rate is representative.[1]

Second, the standard of measurement--thermal content--used in making price indexes does not account for the fact that fuels are not perfectly substitutable on a thermal basis. Some fuels--for example, electricity--are purchased for qualities other than their thermal content. And finally, the price index for the transportation sector applies to gasoline only, and excludes other types of energy used in transport, as well as prices of public transportation. Because public transportation in Western Europe and Japan is much cheaper than in the United States, its exclusion means that the transportation element of the index may overemphasize the difference in prices among the United States and other countries.

---

[1]Electricity rates in Japan appear to be much flatter than those in other countries.

Despite the problems of price measurement, some conclusions may safely be drawn from figure 4-1 and table 4-2. The first is that U.S. energy prices from 1953 to 1976 were consistently lower than those of the other countries. And second, there is some tendency for prices in European countries to draw together over the period, reflecting the increasingly uniform composition of fuel supplies. By 1973, Western European prices averaged about 80 percent higher than those of the United States.

Another aspect of the comparative movement in prices over the period was the stability of average U.S. energy prices up to 1973. In real terms, they drifted gradually downward. But the experience in Western Europe was very different. Exposed to the international oil market, prices showed much greater volatility. In the main, however, the trend in energy prices fell sharply downward until 1970. This occurred because of the fall in real prices of petroleum and electricity and, in the case of the Netherlands, also because of the Groningen natural gas field, which came onstream in the early 1960s.

The general decline in real energy prices over these years would be expected to lead to rising energy consumption. The sharp increases in oil prices, which took place in Europe from time to time in the 1960s, connected with Middle Eastern crises, may not have had much effect in dampening consumption, as it was widely believed at the time (and correctly so as it turned out) that such increases would prove to be transitory.

The 1974 increase in oil prices led to a sharp increase in the level of real energy prices in all the countries we studied. Two

Figure 4-1. Real prices of energy, 1953-76

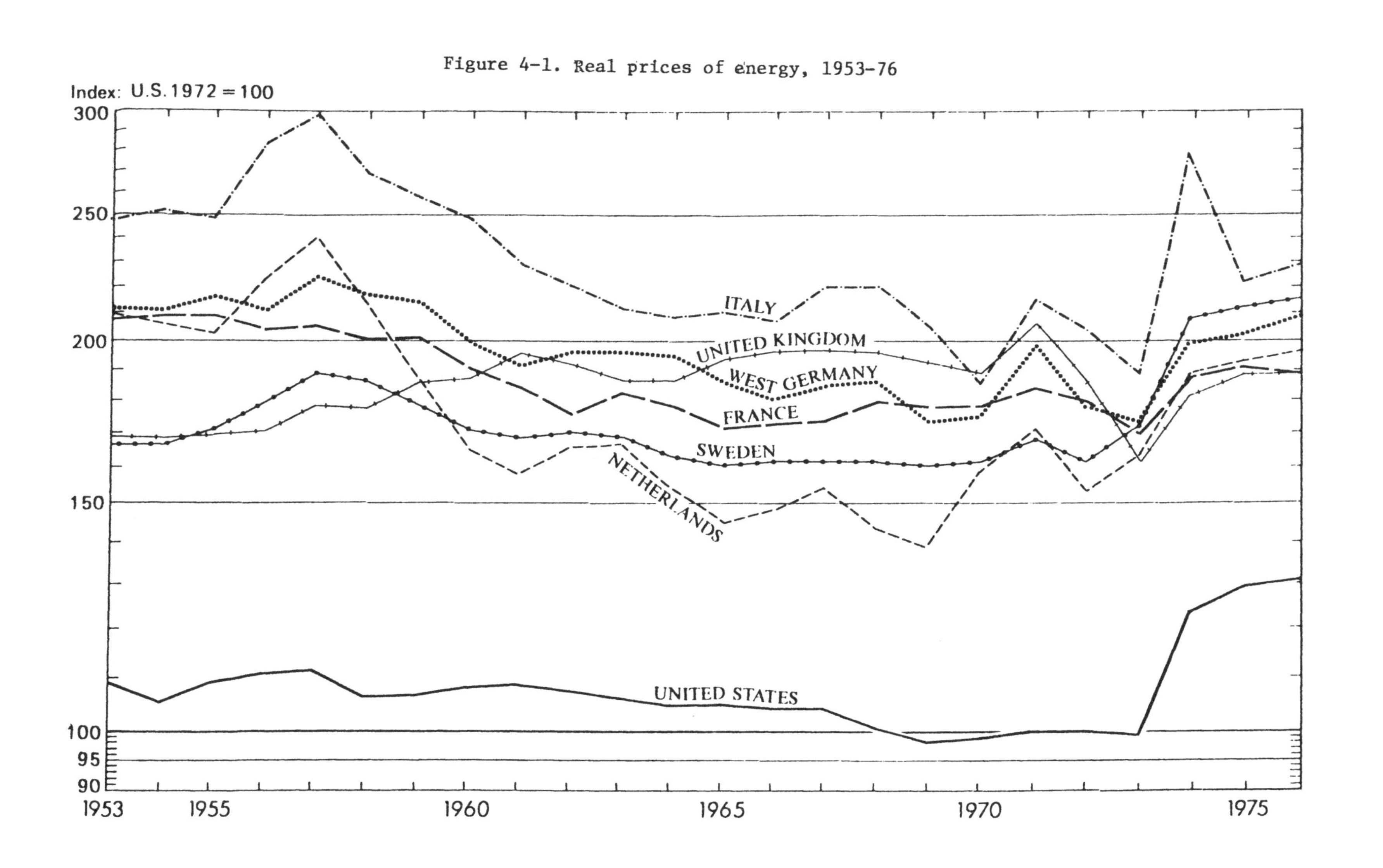

Table 4-2. Indexes of Real Energy Prices Paid By Final Consumers (U.S. 1972=100)

Country	1953	1960	1970	1972	1973	1974	1976
United States	109	108	99	100	99	123	131
Canada	157	154	136	138	133	147	151
France	207	190	178	180	170	187	188
Germany	211	200	175	178	174	199	209
Italy	247	248	186	204	189	219	229
The Netherlands	209	165	160	154	164	188	196
United Kingdom	168	187	189	187	163	181	188
Sweden	167	170	162	162	173	207	215
Japan	269	267	204	211	196	229	272

Source: Based on Joy Dunkerley, Jack Alterman, and John J. Schanz, Jr., Trends in Energy Use in Industrial Societies, EPRI Research Project 864-1 (Palo Alto, CA., Electric Power Research Institute, 1980)

factors are evident about this increase. First, it was of substantially less magnitude than the rise in crude oil prices. The fourfold increase in crude oil prices was diluted at the consumer level by the large energy-processing margins, taxes, inflation, and consumption of other fuels whose prices rose by much less. Typically, the rise in average energy prices between 1973 and 1974 was about 10 to 20 percent.

The second point is that the rate of increase in real energy prices in 1974 was very similar in most countries despite considerable differences in the structure of energy economies and variations in price regimes. For example, the rise in price of energy in the United States was similar to the rise in Japan. To some extent, this similarity is coincidental or, at least, a product of offsetting factors which happened to give the same net figure. Price control, and the higher share of non-oil-fuels in the United States would tend to moderate the increase in total energy prices, whereas the lower tax margin and generally lower inflation rates would tend to increase the U.S. energy price relative to that of other countries. For Japan, the heavy concentration on oil sold at international prices in the fuel supply would increase overall energy prices, but the higher tax margin and inflation rate would modify the increase.

From 1974 to 1978, experience in the countries studied was mixed. High rates of domestic inflation and--in Germany and Japan--the significant appreciation of their currencies against the U.S. dollar moderated the increase in energy prices, often considerably. By the beginning of 1978, for example, prices of gasoline in the Netherlands, Sweden, the United Kingdom, and the United States were below what they had been in 1972, before the OPEC price rise, and in the other countries prices rose typically by about 10 percent (3). The prices of fuel and electricity purchased by households, less affected by taxes than gasoline, rose somewhat more between 1972 and the beginning of 1978. But here again, the increase typically ranged between 10 and 20 percent. That is to say, the sharp rise in crude oil prices in 1974, and the subsequent upward adjustments between 1974 and 1978 that attracted so much attention, resulted in rather modest increases in the real prices of energy paid by the residential consumer. Prices paid by industrial consumers, however, have risen much more strongly. The recent (1978 and 1979) increases in petroleum prices will put further upward pressure on prices.

Geography and Energy Consumption

The effect of energy prices and incomes on energy consumption will be modified by a whole series of factors specific to individual countries. Some of these stem from differences in institutional arrangements, while others are affected by the physical and geographic differences of the individual country. Climate, because of the heating load, is of particular importance. Our countries have considerably different climates. This is illustrated in table 4-3, which gives degree days for each of our countries, measuring the cumulative number of degrees per day in which the actual temperature falls below a threshold temperature at which it is assumed that heating is necessary. The United States is one of the countries having the least number of degree-days, that is, it has one of the mildest winter temperatures. Note that this

Table 4-3. Degree-Day Data 1972

(centigrade degree-days on 18°C threshold)

Country	Degree-days	Country	Degree-days
United States	2700	The Netherlands	3290
Canada	4200	United Kingdom	2840
France	2700	Sweden	4300
Germany	3500	Japan	2400
Italy	2100		

Source: J. Darmstadter, J. Dunkerley, and J. Alterman, *How Industrial Societies Use Energy: A Comparative Analysis* (Baltimore, Md., Johns Hopkins University Press for Resources for the Future, 1977).

results from an average of high degree-days for the North and low degree-days for the South. Only Italy and Japan have fewer degree-days. The other countries have substantially more, especially Canada and Sweden, which have exceptionally cold winters.

Although the coldness of the winter leads to higher consumption because of heating needs, it can also lead to different housing designs and heating systems. In this way, a better insulated house and an efficient heating system can more than compensate for the increased fuel needs. Consumption of heating fuels per household in Sweden, for example, is reported to be lower than in the United States per degree-day because of better insulation and better heating equipment (4).

Just as climate varies from country to country, so it does from year to year, although the variation in degree-days is much greater among countries than over time. A particularly harsh winter, such as those experienced in Europe in 1962 and 1963, might increase degree-days by 20 percent, in contrast to the 100 percent difference that exists among countries. Because only 20 percent of total energy consumption--that devoted to space heating--will be directly affected, it is unlikely that a harsh winter will in itself push up energy consumption by more than 4 percent.

With the advent of air conditioning, summer climate also is relevant to energy consumption. This is much more difficult to measure. Average summer temperatures, because of the variation within one day, give a poor indication of cooling degree-days. We have direct evidence of the amount of energy used in air conditioning in the United States--about 4 percent of the total in 1972--but we do not have this information for other countries. This omission is unlikely to be of major importance as most of the other countries have cooler summers than the United States.

The geographic nature of a country will also affect energy consumption through population density, its distribution between urban and rural areas, and the pattern of settlement generally. The United States and Canada have a very low population density (see table 4-4) as compared with European countries.

In itself, this would imply a larger energy consumption associated with greater distance between cities and the need for increased energy in freight transport. But overall density alone is not an adequate measure of settlement patterns. Sweden, for example, has a low overall density, yet the population is concentrated in a small part of the country yielding high urban densities, a pattern more typical of Europe rather than North America. Ideally, to measure differences in urban densities over time and among countries, we need data on urban densities within the major cities rather than nationwide averages. Available data are given in table 4-4, but it has not been possible to assemble satisfactory data for all countries. In particular, we do not have data illustrating the more suburbanized pattern of settlement in the United States as compared with Europe, which leads to greater reliance on the automobile. Nor has it been possible to assemble data illustrating this process over time for all our countries.

Table 4-4. Selected Data on Land Size and Density 1972

Country	Land area (million square miles)	Population density per square mile	
		Entire country	Eight most populous areas
United States	3.60	58	9,362
Canada	3.64	6	n.a.
France	0.21	245	13,431
Germany	0.10	643	12,709
Italy	0.12	468	n.a.
The Netherlands	0.02	845	n.a.
United Kingdom	0.09	593	10,678
Sweden	0.17	47	n.a.
Japan	0.14	733	18,689

Source: J. Darmstadter, J. Dunkerley, and J. Alterman, *How Industrial Societies Use Energy: A Comparative Analysis* (Baltimore Md., Johns Hopkins University Press for Resources for the Future, 1977).

n.a = not available

Differences in Tastes

Differences in tastes over time, and among countries, can also affect energy-consumption habits. It can be argued that many of the "differences in tastes" can, in turn, be attributed to traditionally low or high prices. But, even if prices were suddenly to change--the more likely outcome would be an upward swing--habits well established in custom and history are unlikely to change as rapidly. Conceptions of what is a luxury and what is a necessity die hard. Take house heating, for example. Standards of heating comfort have probably always been higher in the United States than in European countries as, for example, is shown by a quotation describing conditions around 1850:

> All cabin dwellers gloried in the warmth of their fireplaces, exploiting their world of surplus trees where a poor man, even a plantation slave, could burn bigger fires than most noblemen in Europe...
>
> In the dead of winter, a family kept warm, not by buying "sich uppish notions" as blankets, but by putting more wood on the fire and sleeping in their clothes....The kind of hospitable settler who burned a whole log in order to boil a kettle of tea didn't consider his fire psychologically good until he had crammed a quarter of a cord into a space eight feet wide and four feet deep and had a small-scale forest fire roaring in front of him. If the fire was too hot, he left the door open, but fire he would have, if only to brighten up the dark end of the house.(5)

Even if U.S. energy prices were to approach European levels, it is unlikely that this long history of heating comfort would be abandoned. Differences in tastes are likely to persist even in the unlikely event that all other things become equal.

Differences in standards of heating comfort, so noticeable between countries, also change over time, as testified by the rapid growth of central heating installations in Europe over the past twenty years (6,7). These new standards are also likely to be resistant to price rises.

This chapter has surveyed some of the main factors assumed to affect energy consumption--levels of income, energy prices, geography, and tastes. Our survey suggests that the United States would be expected to have relatively high levels of energy consumption because of high incomes, relatively low energy prices, and sparse population patterns.

The same applies for Canada except that the rigorous Canadian winter is an added stimulus to energy consumption. The same line of argument suggests, as is the case, that a country like Japan with low incomes (compared with the other countries considered here), high energy prices, dense population, and with a mild climate, would have a relatively low level of energy consumption. Over time, of course, in all countries, the sharp rise in incomes and decline in prices (at least until 1973) would be expected to lead--as it has--to rapid increases in energy consumption. Similarly, the rise in energy prices and stagnating growth rates, which occurred after 1973 and up to 1976, would be expected to lead to declining energy consumption. The next step is to try to quantify these relationships more precisely.

REFERENCES

1. Battelle Columbus Laboratories, Potential for Energy Conservation in the Steel Industry (Columbus, Ohio, Battelle Columbus Laboratories, 1975).

2. Gordian Associates Inc., Industrial International Data Base Pilot Study: The Cement Industry (New York, Committee on the Challenges of Modern Society of the North American Council, 1976).

3. International Energy Agency/Organisation for Economic Co-operation and Development, Energy Conservation in the International Energy Agency (Paris, IEA/OECD, 1979).

4. L. Schipper and A. J. Lichtenberg, "Efficient Energy Use and Well-Being: The Swedish Example," Science, vol. 194 (Dec. 3, 1976) pp. 1001-1013.

5. S. Schurr and B. C. Netschert, Energy in the American Economy 1850-1975--Its History and Prospects (Baltimore, Md., Johns Hopkins University Press for Resources for the Future, 1960) pp. 44-5.

6. National Economic Development Office, Energy Conservation in the U.K. (London, HMSO, 1974).

7. J. A. Over, ed., Energy Conservation: Ways and Means (Amsterdam, Future Shape of Technology Foundation, 1974).

Chapter 5

PRICES AND INCOME ELASTICITIES: AGGREGATE ENERGY CONSUMPTION

The previous chapter described the steady rise up to 1973 in per capita energy consumption for all countries, associated with a rise in the Gross Domestic Product (GDP) per capita and a declining trend in the level of real energy prices. In this chapter we use these data to estimate more formally, using a log-linear model,[1] the relationship between energy consumption, income, energy prices, and where possible, geographic characteristics. This is done in two ways: (1) by using data for each country individually over time; and (2) by using data for all countries in all years simultaneously, that is, by "pooling" data.

The pooling of data offers several advantages. It increases the quantity and, particularly, the range of data. Over time, within individual countries, key variables may vary very little (energy prices before 1973 are a good example). It is consequently difficult to assess the influence of such a variable if the movement has been slight, and in one direction. By pooling, the range of price experience--including countries with both high and low prices--is immediately widened. Pooled data on prices include price levels which vary by 100 percent, in place of country-specific data where price variation, even over a period of twenty years, is typically less than 10 percent (at least up to 1973).

---

[1] This model was chosen on the basis of its simplicity and appropriateness for the data available for analysis--income, average energy price, and aggregate energy consumption. For more elaborate treatments, including analysis of interfuel substitution, see two recent studies by Pindyck (1) and Griffin (2).

The case is similar for other variables; weather changes very little from year to year (and, in any event, we found it impossible to measure such changes in practice). By pooling, it is possible to have data applying to very different weather conditions--degree-days for our coldest countries are over 100 percent higher than those for our warmest countries.

Furthermore, by pooling data one avoids the tendency, present in many economic time-series, for all variables to move together. This feature is particularly noticeable in energy. The typical experience of most countries was for incomes to rise and energy prices to fall. These tendencies reinforce each other; both contribute to higher energy consumption. Consequently, it is difficult to disentangle the relative effects of each. The pooling of data from all countries avoids this colinearity because it introduces data from countries without this reinforcing tendency; for example, data from those countries with relatively low incomes and low energy prices. But even pooled data contain some multicolinearity problems inherent in time series. For example, in our group of countries, those with high incomes also tend to have low energy prices and vice versa.

Finally, the pooled cross-sectional data give a better indication of long-term adjustments to higher prices than do time series. The European countries, for example, have experienced prices higher than those of the United States for generations, and their stock of energy-using equipment reflects these traditionally higher prices. As it takes many years to turn over much of the stock of energy-using equipment, a time series, even one stretching over twenty years, does not adequately capture these adjustments (See references (1) and (3) for further development of this argument).

Results for Individual Countries

A series of results, relating to the effect on per capita energy consumption of changes in energy prices and income per capita for each country over varying periods of time is given in table 5-1. Initial results, based on a twenty-three-year period (1953-74) for each country, indicate a very strong relationship between per capita energy consumption and the per capita GDP, both with regard to the size of the coefficient and its statistical significance. In most cases over this period it is more than one, even when the standard error is taken into account. This means that during these years a given increase in the GDP for many of these countries was associated with a greater increase in energy consumption. France, the United Kingdom, and West Germany had income coefficients of less than one, confirming the declining energy/ GDP ratio signaled at the beginning of this study. But Italy and the Netherlands had income elasticities well in excess of one--at the 1.8 level. In the case of Italy, this probably reflects growing industrialization over the period, and, in the case of the Netherlands, the large increase in energy consumption following the exploitation of the Groningen natural gas field. The price elasticities, on the other hand, are low, frequently do not have the right (minus) sign, and are of variable significance.

Table 5-1. Regression Results for Aggregate Energy Consumption (Time Series 1953-74)

Country	Coefficient	Standard deviation	t ratio
United States			
Constant	0.205	0.223	0.920
Income	1.072	0.034	32.408
Price	-0.054	0.105	0.515
	$\bar{R}^2$ = 0.9896	SE of equation = 0.010	
Canada			
Constant	-0.182	0.743	0.245
Income	1.281	0.095	13.518
Price	0.094	0.321	0.293
	$\bar{R}^2$ = 0.9825	SE of equation = 0.019	
France			
Constant	2.224	1.103	2.017
Income	0.704	0.107	6.583
Price	-0.980	0.470	2.083
	$\bar{R}^2$ = 0.8629	SE of equation = 0.047	
Germany			
Constant	0.045	0.504	0.089
Income	0.885	0.055	16.043
Price	-0.002	0.211	0.008
	$\bar{R}^2$ = 0.9741	SE of equation = 0.019	
Italy			
Constant	-0.373	0.540	0.69
Income	1.874	0.100	18.792
Price	-0.026	0.221	0.120
	$\bar{R}^2$ = 0.9665	SE of equation = 0.044	
The Netherlands			
Constant	-0.723	0.191	3.780
Income	1.832	0.151	35.647
Price	0.123	0.079	1.556
	$\bar{R}^2$ = 0.9896	SE of equation = 0.019	

Continued on next page)

Table 5-1 (continued).

Country	Coefficient	Standard deviation	t ratio
United Kingdom			
Constant	0.490	0.211	2.330
Income	0.779	0.040	19.615
Price	-0.155	0.096	1.613
	$\bar{R}^2$ = 0.9550	SE of equation = 0.011	
Sweden			
Constant	0.295	0.298	0.990
Income	1.423	0.036	39.593
Price	-0.255	0.133	1.918
	$\bar{R}^2$ = 0.9869	SE of equation = 0.017	
Japan			
Constant	0.288	0.450	0.506
Income	1.029	0.038	26.944
Price	-0.158	0.186	0.852
	$\bar{R}^2$ = 0.9971	SE of equation = 0.014	

On the basis of these results, it appears that the income level within a country is the major determinant of the level of energy consumption and that prices are of minor importance. But there are several reasons why these results may not give an entirely satisfactory description of the relationship between energy consumption, energy prices, and income in general, and the relationship between energy prices and consumption in particular.

To begin with, the choice of a log-linear model has implications for the results. The model used here imposes restrictions on substitution among fuels and with other goods, keeps income elasticities constant, and imposes a linear relationship between income and price responses. This last restriction means that it is difficult to assume independent estimates of both price and income elasticities in a time-series model. The variable with the highest variance (in this case, income) will dominate.

The minor role played by prices might also be attributed to an additional characteristic of the equation. It can be argued that energy consumption depends not so much on current energy prices but on past energy prices; that decisions involving the increased use of energy are taken before the energy is actually used, on the basis of prices ruling at that earlier time. Energy consumption is, after all, tied closely to the existing capital stock, and adjustment to higher prices in the short run is limited to reductions in the rate of equipment utilization. The long-run adjustment occurs only when it has been possible to replace existing equipment. There is every reason, therefore, to expect a considerable difference between short- and long-run elasticities, and for the long-run elasticity to be substantially higher. In other words, the results reported here may be short-run elasticities which are normally expected to have low values.

In order to test the possibility of lags in consumption decisions, energy prices were lagged by one year, by two years, and by an average of both years. The price elasticities given by this formulation are much improved. Eight out of nine countries show the right sign, and the values increase to an average of about -0.5. That is to say, a 10 percent increase in energy prices is associated with a 5 percent decrease in energy consumption. Again, the income elasticities are well determined. France, West Germany, and the United Kingdom have income elasticities under one, and are joined in this lagged version by the United States, whose income elasticity drops to 0.83 from the 1.07 given by unlagged data.

Finally, as we pointed out in chapter 2, energy consumption data expressed in gross input terms is likely in some circumstances to underestimate the underlying demand for energy services. Results based on useful energy consumption, although for a shorter period (see table 5-2), raise the income elasticities of most European countries, often substantially. All European countries now have income elasticities substantially over one. The price elasticities, though still less well determined than the income elasticities, have improved.

In summary, the results for individual countries furnish some evidence, albeit rather weak evidence, of a negative relationship between price and energy consumption. It is assumed that these elasticities

Table 5-2. Regression Results for Aggregate Energy Consumption (1960-76 Useful Energy)

	Coefficient	Standard deviation	t ratio
United States			
Constant	0.355	0.126	2.189
Income	0.981	0.042	23.420
Price	-0.235	0.063	3.712
	$\bar{R}^2$ = 0.9716	SE of equation = 0.010	
Canada			
Constant	0.650	0.400	1.624
Income	1.181	0.052	22.650
Price	-0.394	0.181	2.182
	$\bar{R}^2$ = 0.9743	SE of equation = 0.015	
France			
Constant	0.643	0.409	1.572
Income	1.189	0.029	40.622
Price	-0.505	0.183	2.763
	$\bar{R}^2$ = 0.9905	SE of equation = 0.011	
Germany			
Constant	0.470	0.234	2.013
Income	1.336	0.034	39.599
Price	-0.411	0.101	4.077
	$\bar{R}^2$ = 0.9907	SE of equation = 0.010	
Italy			
Constant	0.075	0.362	0.207
Income	1.796	0.060	30.084
Price	-0.307	0.152	2.024
	$\bar{R}^2$ = 0.9858	SE of equation = 0.019	
The Netherlands			
Constant	-0.786	0.187	4.206
Income	2.130	0.045	47.186
Price	-0.002	0.088	0.020
	$\bar{R}^2$ = 0.9941	SE of equation = 0.014	

(Continued on next page)

Table 5-2 (Continued)

Country	Coefficient	Standard deviation	t ratio
United Kingdom			
Constant	-0.091	0.533	0.171
Income	1.041	0.109	9.564
Price	-0.083	0.225	0.367
	$\bar{R}^2$ = 0.8660	SE of equation = 0.018	
Sweden			
Constant	0.067	0.161	0.419
Income	1.512	0.055	27.543
Price	-0.291	0.079	3.683
	$\bar{R}^2$ = 0.9828	SE of equation = 0.012	
Japan			
Constant	1.017	0.231	4.399
Income	1.130	0.025	45.661
Price	-0.587	0.096	6.120
	$\bar{R}^2$ = 0.9946	SE of equation = 0.015	

based on time-series data reflect shorter- rather than longer-term adjustments to changes in energy prices.

Results Based on Pooled Data

One reason for the relatively weak price response noted in these countries is that, apart from 1974 and subsequent years, prices changed very little and, on the whole, in the same direction. In order to widen the range of data, therefore, data for all countries for all years were pooled for the same series of equations. A series of results is given in table 5-3.

First, data for all countries for all years (1953-74) are pooled; second, these data are divided into two subperiods (1953-59 and 1960-74);[2] and finally, the 1960-76 series is given in useful energy to identify the underlying demand for energy services.

Let us take the results for the longer period first: The price elasticity, better determined than in the results for individual countries, is about -0-6. The income elasticity is substantially over one, in the 1.2 range.

These regression results cover a long period of time (twenty-one years) in which there were radical changes in the energy economy of almost every nation except the United States. It is therefore of interest to see whether there were any changes in the income and price elasticities within this period--whether the sensitivity of energy consumption

---

[2]The differences between the two periods tested as statistically significant, but more refined analysis could investigate the plausibility of other subperiods.

Table 5-3. Pooled Regression Results: Per Capita Energy Consumption, GDP, and Energy Prices

Regression results		Coefficient	Standard error	t ratio
Pooled data	1953-74			
Constant		1.245	0.216	5.78
Income		1.231	0.052	23.849
Prices		-0.608	0.088	6.922
Pooled data	1953-59			
Constant		1.724	0.600	2.873
Income		1.149	0.151	7.616
Prices		-0.796	0.244	3.262
Pooled data	1960-74			
Constant		0.807	0.219	3.684
Income		1.378	0.061	22.853
Prices		-0.452	0.088	5.154
Pooled data for useful energy	1960-76			
Constant		-0.065	0.241	0 268
Income		1.196	0.054	22.293
Prices		-0.471	0.071	6.670

to price changes was becoming greater or lesser with time. In order to test this, the data were divided into two parts, the first covering 1953-59 and the second, 1960-74. In the later period the income elasticity becomes much higher, and the price elasticity drops sharply from -0.8 to -0.4.[3]

Two other studies (3, 4), using similar analytic procedures to obtain price and income elasticities for sets of industrialized countries, yielded results which can be easily compared with ours (see table 5-4). All studies show a well-determined price elasticity, much higher in the pooled formulation than in the time series.

There are some differences between the results. Those reported in the present study give higher income elasticities and lower price elasticities. To some extent this may be caused by differences in data. Kouris, for example, explains total (and not per capita) unadjusted energy consumption for European Economic Community countries and a different lag structure.

Given the current debate over the efficacy of energy prices as a regulator of energy consumption, the main conclusion to be drawn from these studies is that a well-determined price response does exist, at

[3] Having established the possibility of some changes in price and income elasticities over time, we must then ask whether within those two periods the price and income elasticities were stable or whether they varied considerably around an average value. Kouris (3) finds, for example, some greater stability in coefficients toward the end of this period, 1970. Our data did not entirely support this view, perhaps because the years 1970-74 were included, during which time the European countries experienced considerable price changes.

Table 5-4. Comparison of Income and Price Elasticities

Study	Income	Price	$\bar{R}^2$
Kouris	0.84	-0.768	0.9290
RFF - 1953-74 (pooled data)	1.231	-0.608	0.9165
Lagged Prices (pooled data)	1.204	-0.626	0.9212
RFF - 1960-76 (pooled data)	1.196	-0.471	0.9177
Nordhaus - 1955-72	0.79	-0.85	0.9880

Source: Data for the Kouris study are found in G. Kouris, "The Determinants of Energy Demand in the EEC Area," *Energy Policy* vol. 4, no. 4 (December 1976). Data for the Nordhaus study are found in W. Nordhaus, ed., *The Demand for Energy: An International Perspective*, Proceedings of the Workshop on Energy Demand (Laxenburg, Austria; International Institute for Applied System Analysis, 1975).

least on the basis of pooled cross-sectional data. In other words, the results are consistent with the hypothesis that energy consumption is responsive to energy prices. A major qualification should be made, however. Since these results are based on pooled cross-sectional data, they will tend to reflect a long-term adjustment of elasticity rather than the short to medium term, which is more often the focus of current energy policy.

REFERENCES

1. R. S. Pindyck, The Structure of World Energy Demand (Cambridge, Mass., MIT Press, 1979).

2. J. M. Griffin, Energy Conservation in the OECD: 1980-2000 (Cambridge, Mass., Ballinger, 1979).

3. G. Kouris, "The Determinants of Energy Demand in the EEC Area," Energy Policy vol. 4, no. 4 (December 1976).

4. W. Nordhaus, ed., The Demand for Energy: An International Perspective, Proceedings of the Workshop on Energy Demand (Laxenberg, Austria, International Institute for Applied Systems Analysis, 1975) pp. 511-587.

Chapter 6

PRICE AND INCOME ELASTICITIES: SECTORAL ENERGY CONSUMPTION

In the foregoing chapters, income and price coefficients were based on aggregate energy consumption data. However, price and income coefficients can vary widely between sectors. Insofar as aggregate coefficients are composites of the individual sector coefficients, the aggregate coefficient may change over time, following changes in the sectoral composition of energy consumption. This sectoral analysis, of interest in itself, will also help explain the composition of the aggregate price and income coefficients.

The Residential-Commercial Sector[1]

From some points of view, the inclusion of residential with commercial uses is unfortunate as it makes it difficult to assess whether the sharp rise in consumption for the sector as a whole was caused primarily by its residential or by its commercial components. This is an important consideration because energy consumption in these two subsectors is likely to be affected by different factors. Residential consumption will be influenced by rises in family income, rates of family formation, changes in the housing stock, age structure of the

---

[1]This sector, more correctly described by the OECD as the "other" sector, is a residual sector consisting of energy consumption which could not be assigned to the other more clearly defined sectors. As such, it includes a wide variety of uses such as government at all levels, street lighting, handicrafts, agriculture, and so on.

population, prices of household appliances, and energy prices that differ from those seen in the commercial sector. On the other hand, energy consumption in the commercial sector is likely to be affected more directly by general economic conditions and by structural changes in a nation's economy, such as the move from manufacturing to a service economy. Though it was possible to separate residential from commercial uses in the case studies for individual countries, here we are obliged to analyze the combined sector for the nine countries.

As indicated in chapters 1 and 2, the rise in consumption of energy for residential-commercial purposes was particularly rapid, especially in Europe. This rise was even more pronounced when expressed in terms of useful energy, as, over this period, the more energy-efficient fuels, petroleum and gas, were widely substituted for coal in the European countries.[2]

Since the 1973-74 price rise, experience has shown striking contrast to earlier trends, although for all countries, it has been remarkably similar. In 1974 consumption fell, but by 1976 it had recovered to 1973 levels, and even higher.

Changes in both incomes and prices (see table 6-1) most likely contributed to these trends. Energy prices generally declined relative to other prices. The fall in real prices of energy in the residential

---

[2]In 1960, for all European countries except Sweden, solid fuels were the major source of energy in this sector, typically accounting for about one-half of the total. By 1974, solids diminished to negligible proportions (except in Germany and the United Kingdom), and all European countries relied heavily on petroleum products and, in some cases, natural gas. All countries experienced a rapid increase, typically doubling, in the share of electricity.

Table 6-1. Real Energy Prices in the Residential Sector, for Selected Years

(U.S. 1972=100)

	1960	1966	1972	1973	1974	1976
United States	113	106	100	97	106	112
Canada	128	117	110	113	114	135
France	133	142	141	135	164	160
Germany	123	123	131	141	156	166
Italy	183	152	132	120	143	130
The Netherlands	128	106	104	104	119	126
United Kingdom	172	180	175	165	168	189
Sweden	121	111	113	117	131	132
Japan	298	219	170	160	165	176

Source: Based on Joy Dunkerley, Jack Alterman, and John J. Schanz, Jr., Trends in Energy Use in Industrial Societies, EPRI Research Project 864-1 (Palo Alto, CA., Electric Power Research Institute, 1980).

Note: The above prices, which include taxes, apply to residential purchases only. Prices are indexes of unit costs, that is, the cost of buying a given number of kilocalories of energy. The move in the price index over time will therefore reflect the changing fuel mix in the residential sector. In all countries, the use of electrictity--a more expensive fuel in terms of thermal content--increased, which means that up to 1973 the declining overall price index took place despite an increase in the share of the most expensive fuel on a thermal content basis.

sector up until 1972-73 was particularly marked in Italy and Japan as the price of petroleum products fell. The Netherlands also experienced sharply decreasing prices but for a different reason: during this period natural gas from the Groningen field was being developed and sold at highly advantageous rates. In other countries--France, Germany and the United Kingdom--the move in real prices was less marked. This owed something to the protective policy applied to the coal industry. As imported oil made incursions into the market for coal, both countries took action--obliging power stations to burn coal, imposing excise taxes on home-heating fuels, and so forth--which made the price of energy to the consumer higher than it would have been otherwise.

After the oil price rise in 1974, real energy prices paid by residential consumers in all countries rose, generally by about 20 percent. This rise may not have been sufficient to trigger the anticipated widespread move into new energy-conserving habits within this sector. Furthermore, rapid rates of inflation in subsequent years eroded this rise with the result that by 1978 in several countries the real prices of energy paid by the household sector was no higher than in 1973. The oil price rises of 1979 and 1980 have however, put major pressure on household energy prices recently.

Despite these elements of common experience, there are considerable differences among our countries. Differing climates are particularly important, given the large percentage of total residential-commercial energy consumption that is used in space heating.

Taking these three factors together, it seems that until 1973, the sharp rise in residential-commercial energy consumption was caused by sharp, sustained increases in income, and by declining real energy prices. From 1973 to 1976 the picture is more complex. Energy consumption for residential-commercial use fell initially but recovered later. At the same time, energy prices rose, incomes stagnated, and there were two exceptionally mild winters--all factors which would be expected to lead to a decline in consumption. Consequently, it is difficult to assess which was the dominant factor.

More formally, these relationships have been analyzed by relating per capita energy consumption in the residential-commercial sector to both per capita income and prices and, in the case of pooled data, to degree-days (see table 6-2 and table 6-10). The results indicate a moderately responsive (about -0.5) price effect and very strong income coefficients (about 1.7) in the pooled data.[3]

---

[3]The results of other studies differ somewhat from the results given here. For example, Nordhaus (1) finds, on the basis of pooled data, an income coefficient of 1.08 (a figure lower than ours) and a price elasticity of -0.79 (a much higher figure than ours). Part of this difference may stem from the fact that Nordhaus does not include the United States and Canada, which might add substantially to the income elasticity. Griffin (2), by pooling data for all eighteen OECD countries, finds an income coefficient of 1.295 (again, a figure lower than ours) and a price elasticity of -0.899 (again, a much higher figure). Griffin's income elasticity, without introducing prices, is 1.857, about the same as ours with prices. Thus, the major difference between the Griffin and the RFF cross-sectional results stems from the influence of prices on energy consumption in this sector. These results are more similar to ours for the individual countries. Pindyck (3) reports a price elasticity of aggregate energy use in the long run of about -1.0, and an assumed long-run income elasticity of 1.0.

Table 6-2. Regression Results for Residential-Commercial Energy Consumption (1960-76 Useful Energy)

Country	Coefficient	Standard deviation	t ratio
United States			
Constant	-0.431	0.497	0.867
Income	1.182	0.092	12.794
Price	-0.151	0.224	0.674
	$\bar{R}^2$ = 0.9472	SE of equation = 0.017	
Canada			
Constant	0.493	0.243	2.032
Income	1.292	0.038	33.983
Price	-0.588	0.112	5.224
	$\bar{R}^2$ = 0.9894	SE of equation = 0.011	
France			
Constant	-0.670	0.486	1.379
Income	2.037	0.079	25.938
Price	-0.409	0.240	1.704
	$\bar{R}^2$ = 0.9892	SE of equation = 0.020	
Germany			
Constant	-0.925	0.537	1.72
Income	2.428	0.151	16.055
Price	-0.331	0.279	1.189
	$\bar{R}^2$ = 0.9696	SE of equation = 0.030	
Italy			
Constant	0.011	0.733	0.015
Income	2.174	0.183	11.887
Price	-0.660	0.311	2.120
	$\bar{R}^2$ = 0.9856	SE of equation = 0.027	
The Netherlands			
Constant	-1.324	0.454	2.917
Income	2.715	0.089	30.388
Price	-0.122	0.213	0.575
	$\bar{R}^2$ = 0.9852	SE of equation = 0.028	

(Continued on next page)

Table 6-2 (continued)

Country	Coefficient	Standard deviation	t ratio
United Kingdom			
Constant	-1.140		1.162
Income	1.616	0.109	14.832
Price	-0.006	0.438	0.014
	$\bar{R}^2$ = 0.9317	SE of equation = 0.020	
Sweden			
Constant	0.644	0.439	1.469
Income	1.831	0.089	20.627
Price	-0.921	0.218	4.226
	$\bar{R}^2$ = 0.9638	SE of equation = 0.021	
Japan			
Constant	2.821	1.186	2.380
Income	0.757	0.238	3.176
Price	-0.653	0.474	3.489
	$\bar{R}^2$ = 0.9928	SE of equation = 0.023	

The Transportation Sector

The transportation sector comprises both passenger and freight transport whether by road, rail, air, or inland waterway. Passenger transport is associated with levels of consumer expenditure, the stock of cars, type and size of cars, and prices of gasoline. Freight transport is more closely affected by the level of industrial and economic activity and the size of the internal market.

In Europe and Japan there were major modal changes within the sector. In these countries the share of road transport (always the primary mode) within the total increased sharply, typically by 20 or 30 percentage points, while the share of rail transport fell equally heavily (see table 6-3). The sharp rise in road transport reflects the spread of car ownership and rising per capita gasoline consumption, both of which, however, still remain much higher in the United States than elsewhere (see table 6-4). In contrast, in the United States there was little change in the modal mix within the transport sector, though in both areas air transport steadily increased its still relatively small share of the total.

The stability of the U.S. modal composition of energy consumption in the transportation sector, combined with the large changes in the other countries, meant that, by 1976, patterns of consumption within the transport sector were very similar in North America and Europe. Road transport was the dominant mode, accounting for about 80 percent of the total in all countries, followed by air (7 to 14 percent) and rail (4 to 6 percent).

Table 6-3. Change in Percentage Share of Different Modes in Transport Energy Consumption 1960-76

	United States	France	Germany	Italy	United Kingdom	Japan
Road	+1.9	+27.2	+29.6	+15.1	+26.0	+22.6
Rail	-2.9	-28.1	-32.8	-13.5	-28.5	-35.5
Air	+1.0	+2.9	+5.8	+1.6	+4.9	+1.6
Inland navigation	n.a.	-2.0	-2.6	-3.2	-2.4	+11.3

Source: International Energy Agency/Organisation for Economic Co-operation and Development, Energy Balances of OECD Countries, various issues (Paris, OECD).

n.a. = not available

Table 6-4. Car Ownership and Gasoline Consumption

Country	Car ownership				Gasoline consumption	
	1960		1974		1960	1976
	Cars/ 1,000 pop.	U.S.= 100	Cars/ 1,000 pop.	U.S.= 100	kilos per capita	kilos per capita
United States	344	100	492	100	915	1,401
Canada	237	69	377	77	659	1,130
France	133	37	269	58	108	331
Germany	92	27	272	55	97	342
Italy	48	14	258	52	50	193
The Netherlands	53	15	254	52	104	266
United Kingdom	113	33	246	50	139	299
Sweden	173	50	323	66	193	396
Japan	7	2	144	29	44	193

Source: Car ownership data are derived from United Nations, Statistical Yearbook, various issues. Gasoline consumption data are based on Organisation for Economic Co-operation and Development, Energy Statistics, various issues (Paris, OECD).

In Europe, the sharp decline in rail's share indicates not only a movement of passengers and freight out of the rail sector, but also a sharp improvement in efficiency of rail operations (measured by energy consumed per passenger-mile or tons per mile). This was largely the result of increased use of diesel-powered equipment, a process which had been virtually completed by 1960 in the United States.[4]

Despite the growing similarity in modal distribution of transport energy consumption, the United States still uses much more energy in transportation. But, as in other sectors, the gap is narrowing: at the beginning of the period, European consumption per capita was only 20 percent of the U.S. level. By 1976, however, consumption by European countries rose to about 29 percent of that of the United States. Consumption fell or stagnated in all countries in 1974, and although some growth took place in the two years since then, the rate of increase has been much less rapid than in previous years.

As this sector is virtually completely dependent on petroleum products, with but a minimal possibility of fuel switching, changes in oil prices are clearly of critical importance. All countries experienced a strong decline in real gasoline prices in the 1960s and early 1970s (see table 6-5). In 1974, following the price rise in crude oil, they increased an average of 20 percent. Despite the radical

---

[4]Another contributing factor was probably the closing down of routes and lines which took place in many European countries during this period. The elimination of lines that carry few passengers automatically increases "efficiency" by lowering the energy use per passenger-mile on the better traveled lines that remain. There may still be efficiency gains to be achieved from this factor in the future.

Table 6-5. Real Prices of Gasoline, 1960-76

(U.S. 1972=100)

Country	1960	1966	1972	1973	1974	1976
United States	121	117	100	100	125	118
Canada	129	124	114	113	122	115
France	298	270	235	221	273	242
Germany	266	194	175	172	193	193
Italy	436	286	314	283	298	215
The Netherlands	269	246	232	220	246	230
United Kingdom	255	251	230	221	271	190
Sweden	263	239	201	203	219	193
Japan	241	243	201	205	203	165

Source: Based on Joy Dunkerley, Jack Alterman, and John J. Schanz, Jr., Trends in Energy Use in Industrial Societies, EPRI Research Project 864-1 (Palo Alto, CA., Electric Power Research Institute, 1980).

Note: Because of the predominance of gasoline in total energy consumption, gasoline prices have been used as a proxy for all prices facing the transportation sector.

change from past experience which this increase represented, it is surprising, at first sight at least, that such a sharp (fourfold) increase in crude oil prices resulted in such a modest rise in product prices. The reasons for this are, broadly, that refining and distribution costs and excise taxes represent a large part of total gasoline costs and, therefore, dilute the effect of the rise in crude oil prices. And second, the sharp rates of inflation in many countries in these years absorbed an important part of the increase in gasoline prices. This is particularly noticeable in the years after 1974 and up to early 1978, when gasoline prices fell, often heavily, in real terms, primarily because of the sharp rise in the consumer price index which was used to deflate gasoline prices in order to express them in real terms. In several of our countries, gasoline prices in early 1978 were, in real terms, lower than they had been in 1972, before the fourfold increase in oil prices.

Relating gasoline prices and per capita income to per capita energy consumption in transportation for each of our countries (see table 6-6 and table 6-10) yields a high price elasticity (about -1.0) from the pooled data, though lower from country data, and income elasticities of about 1.0.

A particular note of caution in the interpretation of these elasticities is in order here. It is of arguable value to relate the consumption of energy in the transport sector to gasoline prices and income only. The amount of energy consumed in this sector is also determined by the stock of cars and vehicles and their energy-using characteristics. While it can be said that gasoline prices and per capita income

Table 6-6. Regression Results for Transportation Energy Consumption (1960-76 Useful Energy)

Country	Coefficient	Standard deviation	t ratio
United States			
Constant	1.166	0.302	3.866
Income	1.058	0.070	15.194
Price	-0.003	0.137	0.094
	$\bar{R}^2$ = 0.9421	SE of equation = 0.015	
Canada			
Constant	0.018	0.750	0.023
Income	1.079	0.091	11.873
Price	-0.605	0.340	1.779
	$\bar{R}^2$ = 0.9789	SE of equation = 0.013	
France			
Constant	-1.303	0.153	8.519
Income	1.270	0.026	48.339
Price	-0.183	0.061	3.012
	$\bar{R}^2$ = 0.9943	SE of equation = 0.009	
Germany			
Constant	-0.623	0.366	1.702
Income	1.501	0.104	14.484
Price	-0.535	0.137	3.900
	$\bar{R}^2$ = 0.9893	SE of equation = 0.014	
Italy			
Constant	-1.294	0.313	4.129
Income	1.456	0.100	14.555
Price	-0.180	0.118	1.518
	$\bar{R}^2$ = 0.9547	SE of equation = 0.029	
The Netherlands			
Constant	-1.073	0.846	1.268
Income	1.295	0.109	11.844
Price	-0.246	0.337	0.728
	$\bar{R}^2$ = 0.9614	SE of equation = 0.023	

(Continued on next page)

Table 6-6 (continued)

Country	Coefficient	Standard deviation	t ratio
United Kingdom			
Constant	-2.115	0.263	8.042
Income	1.846	0.084	21.958
Price	-0.077	0.101	0.761
	$\bar{R}^2$ = 0.9734	SE of equation = 0.01	
Sweden			
Constant	-1.726	0.798	2.162
Income	1.176	0.193	6.098
Price	0.044	0.290	0.151
	$\bar{R}^2$ = 0.9239	SE of equation = 0.021	
Japan			
Constant	-1.724	0.254	6.783
Income	1.151	0.042	27.706
Price	0.012	0.102	0.121
	$\bar{R}^2$ = 0.9955	SE of equation = 0.013	

in themselves give a good indication of both the stock of vehicles and their characteristics, it is more satisfactory to have a closer idea of adjustment processes.[5]

Furthermore, it is necessary to take into account changes in the road and highway network over this period. The extensiveness and the nature of this network closely influences the stock of vehicles; the provision of a fast intercity road network will lead to the development of a different type of auto--larger, more comfortable, with lower mileage per gallon of gasoline. Even with high gasoline costs, intercity car travel on improved roads might prove to be cheaper than alternative methods. Similarly, the improvement of access roads to cities can lead to the purchase of cars by those who would otherwise have used public transport for commuting, leading to the loss of revenues to public transportation systems and resulting in their subsequent deterioration. Both seem to have been of significance for energy consumption in the European transport sector in the 1960s. If the highway system is highly influential in determining energy consumption in this sector, it is not, however, necessarily tied to periods of high economic activity or declining gasoline prices, although it may coincide with such periods. Any analysis based solely on gasoline prices and income may not capture these effects and, therefore, should be considered incomplete.

---

[5]The Griffin and Pindyck studies (2, 3) remedy this to some extent by introducing into the analysis of gasoline consumption the stock of cars and their utilization and efficiency characteristics.

The Industrial Sector

The rise in industrial energy consumption in the 1960-76 period was slower than the rise in other sectors, and generally slower than the rise in the GDP and in industrial output. This is, at first sight, surprising; as is common with energy prices generally, prices for energy paid by industrial consumers on the whole either fell or remained stable in real terms up until 1973 (table 6-7).

Insofar as energy is a substitute for labor as an industrial input, then the price of energy in terms of wages fell very sharply indeed. This would normally be expected to lead to a greater than, rather than smaller than, proportional increase in industrial energy consumption as industrial production rose.

But, for two possible reasons, this did not take place. First, in many industries energy is only a small component of total costs and, because of its relative unimportance, may have escaped close economic scrutiny within the industrial sector. Second, even if this were not so, energy is but one factor of production, and its use will depend not only on the price of energy, but on the price and use of other factors of production, in particular, capital and labor.

A series of studies (3, 4, 5, 6) have attempted to model industrial energy consumption in relation to other factors of production in order to test how far energy can be substituted for other factors. All find that energy can be substituted for labor, though there is as yet an unresolved debate with regard to whether energy is a substitute for or a complement to capital. In general, studies based on individual

Table 6-7. Real Energy Prices Paid by Industrial Consumers, 1960-76 (U.S. 1972=100)

Country	1960	1966	1972	1973	1974	1976
United States	101	98	100	100	131	145
Canada	172	155	158	147	169	174
France	197	165	186	167	177	188
Germany	231	205	203	187	221	234
Italy	227	204	206	194	251	276
The Netherlands	163	151	168	188	217	230
United Kingdom	182	193	184	153	171	188
Sweden	181	180	187	202	256	272
Japan	241	225	221	205	254	318

Source: Based on Joy Dunkerley, Jack Alterman, and John J. Schanz, Jr., Trends in Energy Use in Industrial Societies, EPRI Research Project 864-1 (Palo Alto, CA., Electric Power Research Institute, 1980).

countries' time-series data indicate that energy and capital are complements, whereas those based on cross-sectional international data indicate that they are substitutes. Part of this apparent discrepancy may be caused by differences in defining the constituent factors of production, in particular whether three factor inputs (capital, labor, energy) are included, as they usually are, in international analyses, or whether four (capital, labor, energy, and other materials) are included, as they often are in intracountry studies. The inclusion of the fourth factor input has been shown in some circumstances to convert energy-capital subsitutability into energy-capital complementarity. An alternative explanation of the apparent discrepancy in results is that the intracountry analyses yield short-term cross-elasticities, whereas the intercountry analyses capture long-term adjustments. This is a topic which is clearly in need of further research.

In the meantime, what seems to have happened in our countries up to 1973 is that some substitution of capital equipment (and energy) was made for labor in response to rising labor costs, but that this movement was more than offset by technological developments--such as the substitution of more energy efficient new industrial equipment for old. This appears to have been an important influence on industrial energy consumption in the manufacture of iron and steel and cement.

This sort of development is illustrated by the Swedish experience. In that country the price of energy in real terms declined over the period. Industrial wages rose steeply, and a recent study (7) indicates that the user price of capital, though declining in the fifties, rose quite sharply in the sixties. The overall increase in the cost of capital was, however, much less than the rise in industrial wages, so that there is some economic presumption of a move to labor-saving technologies through the substitution of capital and energy. Such a move would tend to increase the energy intensity of industrial production.

At the same time, however, there is evidence of the introduction of cost-saving technology over the period as a whole, so that the output of Swedish industry was increasing more rapidly than inputs into the industrial sector.

Table 6-8 illustrates both points made above. First, inputs of capital and energy were increasing more rapidly than inputs of labor and, second, industrial production was increasing more rapidly than total inputs of capital, labor, and energy. The net result was a rapid rate of growth in energy consumption, but less rapid than the growth in industrial production, resulting in a diminished intensity of energy use relative to industrial production.

Given these complex linkages, about which so little is know, the association of industrial energy consumption with the GDP and energy prices paid by the industry sector offers a very limited approach. It is nonetheless done here to round off the sectoral analysis. The result (see table 6-9 and table 6-10) indicate, in many cases, a rather lower income coefficient and low price elasticities (in the -0.2 range).[6] These low price elasticities are surprising given the much sharp increase in industrial energy prices, compared with the other sectors. The inclusion of more up-to-date data might modify these results.

---

[6]The Nordhaus (1) study gives a price elasticity of -0.52 and an income elasticity of 0.76. The Griffin study (2), on the other hand, results in a higher price elasticity of -0.8.

Table 6-8. Sweden: Annual Change in Industrial Production, Capital Stock, Hours of Work, and Energy Consumption

(percentage)

Year	Industrial production	Capital stock	Hours of work	Energy consumption
1955-60	5.2	4.8	0.2	3.1
1960-65	7.4	5.5	0.6	4.3
1965-70	4.9	4.2	-1.5	4.1
1955-70	5.8	4.8	-0.2	3.8

Source: V. Bergstrom, "Industriell utveckling, industrins kapitalbildning och finanspolitiken," in E. Lundberg and coauthors, *Svensk finanspolitik i teori och praktik*. Lund, 1971, quoted in L. Bergsman, *Energy and Economic Growth in Sweden--An Analysis of Historical Trends and Present Choice* (Stockholm: Economic Research Institute, Stockholm School of Economics, 1977).

Table 6-9. Regression Results for Industrial Energy Consumption (1960-76 Useful Energy)

Country	Coefficient	Standard deviation	t ratio
United States			
Constant	0.765	0.122	6.255
Income	0.865	0.066	13.083
Price	-0.599	0.067	8.989
	$\bar{R}^2$ = 0.9197	SE of equation = 0.014	
Canada			
Constant	0.141	0.516	0.274
Income	1.085	0.079	13.694
Price	-0.324	0.231	1.404
	$\bar{R}^2$ = 0.9238	SE of equation =.0.024	
France			
Constant	0.087	0.2018	0.431
Income	0.717	0.0548	13.101
Price	0.292	0.092	0.318
	$\bar{R}^2$ = 0.9138	SE of equation = 0.020	
Germany			
Constant	0.099	0.284	0.348
Income	0.744	0.052	14.357
Price	-0.256	0.118	0.218
	$\bar{R}^2$ = 0.9415	SE of equation = 0.014	
Italy			
Constant	0.894	0.228	3.915
Income	1.379	0.046	29.778
Price	-0.729	0.099	7.346
	$\bar{R}^2$ = 0.9822	SE of equation = 0.016	
The Netherlands			
Constant	-0.847	0.146	5.813
Income	1.696	0.060	28.203
Price	-0.086	0.073	1.180
	$\bar{R}^2$ = 0.9875	SE of equation = 0.016	

(Continued on next page)

Table 6-9 (continued)

Country	Coefficient	Standard deviation	t ratio
United Kingdom			
Constant	0.400	0.405	0.987
Income	0.334	0.097	3.438
Price	-0.305	0.169	1.810
	$\bar{R}^2$ = 0.5857	SE of equation = 0.016	
Sweden			
Constant	-0.446	0.193	2.314
Income	1.235	0.103	12.036
Price	-0.126	0.100	1.257
	$\bar{R}^2$ = 0.9321	SE of equation = 0.019	
Japan			
Constant	0.045	0.187	4.529
Income	0.996	0.024	42.058
Price	-0.608	0.079	7.658
	$\bar{R}^2$ = 0.9910	SE of equation = 0.016	

Table 6-10. Summary of Regression Results Based on Pooled Data for 1960-76

1960-76	Income elasticity	t ratio	Price elasticity	t ratio	$\bar{R}^2$
Aggregate including degree-days	1.196	22.293	-0.471	6.670	0.9177
Residential-commercial including degree-days	1.725	22.756	-0.596	5.521	0.9095
Industry	1.132	26.513	-0.198	3.534	0.8680
Transport	1.094	18.889	-0.996	17.570	0.9290

The above analysis of industrial energy consumption does not include energy consumed by the petrochemical industry and other nonenergy uses. Energy for these uses rose sharply, particularly outside North America. Consumption over the relatively short period 1960-76, for example, frequently increased four- or fivefold, and in the Netherlands and Japan by twenty- or thirtyfold. This rapid increase came to an end in 1974 and, for most countries, consumption has continued to decline since. Clearly, the petrochemical and related industries have been among the hardest hit by the crude oil price rise and the subsequent recession.

The sectoral analysis confirms that useful energy consumption is significantly influenced by the level of energy prices. In all sectors higher prices have been associated with declining consumption--that is, the price elasticity has a negative sign. The price elasticities do, however, vary from sector to sector, being particularly high in the transport sector. These results (given in table 6-10) are based on pooled data and are therefore assumed to apply to long-run adjustment processes. The associated income elasticities are high, particularly in residential-commercial uses (invariably over one).

REFERENCES

1. W. Nordhaus, ed. The Demand for Energy: An International Perspective, Proceedings of the Workshop on Energy Demand (Laxenburg, Austria: International Institute for Applied Systems Analysis, 1975) pp. 511-587.

2. J. M. Griffin. Energy Conservation in the OECD: 1980-2000 (Cambridge, Mass.: Ballinger, 1979).

3. R. S. Pindyck. The Structure of World Energy Demand (Cambridge, Mass.: MIT Press, 1979).

4. E. Berndt and D. Wood. "Technology, Prices and the Derived Demand for Energy," Review of Economics and Statistics, August 1975.

5. E. Hudson and S. W. Jorgenson. "U.S. Energy Policy and Economic Growth 1975-2000," Bell Journal of Economics and Management Science, Autumn 1974.

6. J. M. Griffin and P. R. Gregory. "An Intercountry Translog Model of Energy Substitution Responses," American Economic Review, December 1976.

7. V. Bergstrom, "Industriell utveckling, industrins Kapitalbildning och finanspolitiken," in E. Lundberg and coauthors, Svensk finanspolitik i teori och Praktik. Lund, 1971, quoted in L. Bergman. "Energy and Economic Growth in Sweden--An Analysis of HIstorical Trends and Present Choice" (Stockholm, Sweden: Economic Research Institute, Stockholm School of Economics, 1977).

Chapter 7

ASSESSMENT AND USE OF RESULTS

It is necessary to proceed cautiously in interpreting the results of our study because a number of measurement problems exist. Five will be described here.

First, the data upon which the studies are based--energy consumption, energy prices, and output--are highly approximate and are therefore likely to be measured with substantial error. This point has already been made in connection with data as they were introduced, but it bears repeating. The data are difficult to measure accurately over time and even more difficult to compare from country to country. They cover a period, for the European countries and Japan at least, which saw radical changes in two of the major variables, energy prices and the composition of fuel supplies. A further complication with the price variable is that when prices did rise, such as during the Suez crises of 1956 and 1967, there was among consumers every expectation that the price rise would not be permanent. If price rises are thought to be temporary, they will not lead consumers to purchase more energy-efficient equipment.

Second, the model specification we use here, which relies strongly on income and energy prices as major independent variables, may not be a correct description of reality. It obscures the more direct relationship between the nature and utilization of the capital stock and energy consumption. Though it can be argued that differences in levels of the Gross Domestic Product (GNP) and energy prices over time and between countries in themselves provide a reasonable indication of capital stock characteristics, as long as the relationship is indirect we must retain some skepticism about the validity of results based on the relationship among energy consumption, energy prices, and income only--especially for the future.

Third, many of our results--and the best determined price elasticities--are based on the "pooling" of data for different countries. This operation has the certain danger of incorporating additional and unknown characteristics of the different countries over and above the variables pooled. Or, to put it another way, important variables may have been omitted from the model specifications. Thus, high energy prices might incorporate differences in taste, and differences in views with regard to personal mobility, which might remain or prove very resilient to change even if prices were to change.

Fourth, as energy consumption depends so closely on the stock of capital, the question of long- and short-run elasticities becomes important. Short-run changes in energy consumption are those which can be achieved through the less intensive use of existing equipment and are consequently somewhat limited. Long-run changes are those which occur when the whole capital stock has had time to adjust to the new conditions and have therefore the possibility of being much larger than those for the short run. As we have indicated, experience of the individual countries over time at the level of aggregation used in this study is not considered to yield satisfactory long-run elasticities. Comparisons between countries may give a better indication of longer-term elasticities as the capital stock of a country with traditionally high energy prices is more likely to have adjusted to these prices. But, as noted above, the interpretation of pooled results is itself open to some ambiguity.

Even if long-run elasticities are correctly identified, they would be expected to involve a long adjustment period. Energy consumption in the past was undoubtedly encouraged by low prices stretching over a long period of time. A stock of energy-using appliances was built up in response to these slowly declining prices. These appliances will not immediately be scrapped because of the price rise which has occurred since 1974. Results based on past experience cannot be expected to be immediately reversible in the future. The time taken for long-term adjustment will, of course, vary from sector to sector--Pindyck (1), for example, estimates the half-life for residential heating equipment to be between six and nine years, and that for industrial equipment, three to fifteen years. These periods tend to be longer than the target periods associated with current policies--such as the U.S. intention to reduce oil import dependency by the mid-1980s.

And, finally, caution is indicated in using these results for forecasting. The period covered was a period of rising income and declining energy prices--both factors which would reinforce each other in leading to increased energy consumption. Experience based on such a period, or on the cross-sectional comparison of countries where, again, high income and low energy prices are highly correlated, may not be a good indication of future conditions when either one of these variables is subject to change. Price elasticities based on the experience of past years may not transfer easily to the future.

With these qualifications in mind, we use our results to assess how much of the difference in useful energy consumption is based upon differences in prices and how much is related to differences in income. We take first the difference in energy consumption, as measured in per capita terms among countries. Because our primary interest is the United States, we have taken the U.S. equation and paired it, in turn, with the price and income data of each other country for the year 1972. This procedure gives us:

1. Useful energy consumption per capita, given U.S. income and U.S. energy prices
2. Useful energy consumption per capita, given the income and prices of the other country
3. Useful energy consumption per capita, given U.S. income and energy prices of the other country.

The total difference in energy consumption (item 1 minus item 2) can then be subdivided into that part of the difference caused by prices (item 1 minus item 3) and the remainder which is attributed to differences in income (item 3 minus item 2).[1] The results are given in table 7-1.

This table shows that income is responsible for most of the difference, but prices are also of importance, accounting typically for 20 to 30 percent of the differences in per capita energy consumption among countries. Repeating the same exercise, using the pooled data equation with its larger price elasticity, increases the part attributed to prices to almost 50 percent.

Although chapters 5 and 6 have been devoted to the explanation of per capita consumption over time and among countries, the original focus was the energy/output ratio. In particular, we began by asking, "What causes energy/

---

[1]This treatment contains some ambiguity. Thus, an alternative formulation would be to define another pair: item 4, U.S. prices and other country incomes. Then item 1 minus item 4 yields an estimated effect attributable to income, and item 4 minus item 2 yields the alternative price effect. This formulation would give different results--the price effect would be smaller--but would not affect the main conclusions.

Table 7-1. Energy Consumption Per Capita in 1972, Using U.S., Other Countries' Prices or Income, or Both, 1970-76

(in toe/capita)

Countries	U.S. prices, U.S. income (1)	Other prices, other income (2)	Other prices, U.S. income (3)	Total difference (col. 1 minus 2)	Difference caused by prices (col. 1 minus 3)	Difference caused by income (col. 3 minus 2)
U.S. - Canada						
U.S. equation	4.191	3.271	3.885	0.920	0.306	0.615
Pooled equation	4.041	2.815	3.472	1.227	0.569	0.658
U.S. - France						
U.S. equation	4.191	2.711	3.650	1.480	0.541	0.939
Pooled equation	4.041	2.132	3.064	1.909	0.977	0.932
U.S. -Germany						
U.S. equation	4.191	2.605	3.660	1.586	0.531	1.055
Pooled equation	4.041	2.035	3.080	2.006	0.961	1.045
U.S . - Italy						
U.S. equation	4.191	1.665	3.544	2.527	0.646	1.880
Pooled equation	4.041	1.149	2.889	2.892	1.253	1.739
U.S. - The Netherlands						
U.S . equation	4.191	2.488	3.787	1.703	0.404	1.299
Pooled equation	4.041	1.976	3.298	2.065	0.744	1.322
U.S. - United Kingdom						
U.S. equation	4.191	2.200	3.618	1.991	0.573	1.417
Pooled equation	4.041	1.641	3.009	2.400	1.032	1.368
U.S. - Sweden						
U.S. equation	4.191	3.382	3.804	0.809	0.387	0.442
Pooled equation	4.041	2.883	3.328	1.158	0.713	0.445
U.S. - Japan						
U.S. equation	4.191	2.153	3.516	2.038	0.674	1.363
Pooled equation	4.041	1.563	2.843	2.478	1.198	1.280

ratios to change over time and to differ between countries?" By taking the energy/GDP ratio as the measure of consumption we are, of course, already incorporating a major part of the income effect in the measure to be explained; or, to put it another way, we are excluding a major part of the income effect from the explanatory factors. Only when the income coefficient is substantially over unity can we say that the income effect over and above the unitary coefficient has contributed to changes in energy/GDP ratios over time and in differences among countries.

As we have seen, for the individual countries over time, the income coefficients were in excess of one, so that rising income played some part in the total useful energy/GDP ratios of the European countries in the 1960-73 period. Declining real prices also contributed, but the generally low values of the price coefficients and, in many cases, their uncertain significance, imply that over time income was the major force behind changes in useful energy/GDP ratios. After 1973, however, there is evidence of the greater influence of prices, but here again, it is difficult to be specific, as the fall in energy/GDP ratios which took place after the OPEC price increase also owed something to industrial recession. On the other hand, the differences among countries in energy/GDP ratios is largely caused by differing prices--perhaps by as much as 60 percent.

In general, therefore, the convergence between the U.S. useful energy/GDP ratio and that of other countries shows the influence of the higher income elasticity in European countries, but the considerable gap that remains stems from higher European energy prices. This implies that pricing policies can be an effective way of moderating energy demand, though the time period for them to become fully effective should not be underestimated.

REFERENCE

[1]R. S. Pindyck, The Structure of World Energy Demand (Cambridge, Mass., MIT Press, 1979).

Chapter 8

ENERGY POLICY RESPONSES

Policies with regard to energy can potentially have a major influence on how much energy is consumed. In this chapter we describe the energy policies followed by our countries over this period, particularly in the post-1970 period. We then assess the effect of these policies.

Policies affecting energy are of two kinds: those specifically directed to energy consumption or supplies, and those dealing with the overall management of the economy or of constituent sectors of the economy which may have major implications for energy. Policies of the first sort are, by definition, motivated by concern with energy problems and are expected to have planned effects on the problem. An example of this type of policy is an oil import quota. Policies of the second sort, however, are motivated by wider concerns, and their effect on energy may not be foreseen. Examples of this second type of policy are regulations applying to the transport sector, or to housing. The two types of policies may reinforce or may offset each other.

With regard to policies affecting the overall management of the economy, our countries, in the twenty years or so preceding the 1973-74 oil price rise, followed macroeconomic policies which were associated with a rapid increase in energy consumption. Thus the main thrust of these policies was to promote rapid economic growth. Industrial production was encouraged either directly, by tax incentives to investment, or indirectly by removing obstacles such as controls which had grown up during the war economy. Major changes in transport and road-building policies were instituted. Housing policy was geared to the rapid renewal and expansion of the housing stock, which had deteriorated during the war years.

The combined effect of these actions resulted in a rapid increase in energy consumption; nonetheless, some of these policies did in practice have some energy-conservation effect, although this was not the particular intention. Thus, high industrial activity and investment incentives may have led to a more up-to-date and energy-efficient industrial capital stock and the rapid rate of building led to an increasing proportion of new houses, which, if inadequately insulated by today's standards, may have met higher standards than those of the houses they replaced.

The rapid increase in energy consumption in these years set the stage for policies devoted more specifically to energy. Thus the general energy policy concerns of industrialized countries in this period were implicitly or otherwise, "the assurance of adequate and reasonably secure supplies of the various forms of energy needed to sustain economic growth" and "the assurance of reasonable energy prices and the encouragement of reduction of the cost of energy to the community as a whole and to the individual consumer" (1).[1] Though such policy statements closely fit the experience of these years, virtually none of the countries possessed at that time a comprehensive energy policy covering all parts of the energy sector. Measures taken were rather of an *ad hoc* nature whose relationship to the broader picture of energy policy is more obvious in retrospect than it was at the time.

For the countries poor in energy resources, access to adequate supplies at reasonable energy prices was achieved by rapidly rising petroleum imports. This occurred spontaneously, encouraged by the price and other advantages of

---

[1] As is seen in reference 1, there were other considerations which could have considerable impact on the energy sector, such as the budgetary, balance of payments, resource and public health questions, and regional and employment problems. In retrospect, however, the assurance of adequate supplies at low prices appears to have been the dominant concern.

oil over its main competitor, coal, and by easy supply conditions (2). The role of government policy in this development was to permit and facilitate the rapid increase in oil imports despite growing dependence on a single area (the Middle East). In some countries, however--particularly France--encouragement was given to diversifying sources of oil imports and to setting up national oil enterprises in order to provide greater security of supply.

For the countries with energy resources--the United States with coal and oil, and the United Kingdom and Germany with coal--the influx of cheap Middle Eastern oil competing with domestic energy supplies caused many problems. All three countries moved to protect their domestic industries: the United States established an oil import quota, and Germany and the United Kingdom offered a wide variety of subsidies to the coal industry and taxed those petroleum products competing with coal (2). Thus these countries tempered their policy of reducing the cost of energy to the individual consumer by the claims of overriding national concerns.

A policy common to all countries (particularly the European countries) was the imposition of heavy rates of taxation on gasoline. It is important to note, given our subsequent interest in energy conservation, that the motive behind these taxes was primarily fiscal rather than conservationist.

Since the major increase in oil prices in 1973-74 and the partial embargo there have been important changes in energy policy perceptions among all countries included in our study and in other countries too. In general there has been a growing realization that energy must be treated as an integral sector--not as a series of isolated fuel-supply issues--and must include conservation as well as increased supply. Second, the dangers of undue dependence on imported oil have been illustrated dramatically, and most countries plan to protect themselves against future supply disruption by reducing their import dependence and increasing their strategic stockpiles.

Part of this protection is to be assured by memberships in the International Energy Agency. All of the countries included in this study are members, with the exception of France. The two main aspects of the International Energy Program are an oil-allocation scheme in times of emergency and a provision for long-term cooperation on energy.[2] Under the emergency allocation scheme, each country is obliged initially to maintain oil stock piles sufficient to sustain consumption for a period of sixty days with no imports. And each country must have ready a program of demand restraint measures to be implemented in time of emergency.[3] For a detailed analysis of how this allocation scheme would work out under certain assumptions, see reference 4. By subscribing to this program, the member countries hope to minimize the disruption to the group as a whole that could be caused by a sudden shortfall in oil supplies.

The program dealing with long-term cooperation on energy is aimed at a longer-term reduction in dependence on imported oil. It includes cooperative programs in energy conservation, the development of alternative energy sources, energy research and development, and uranium enrichment.

---

[2]Other parts of the program include an information system on the world oil market, consultation with oil companies, and relations with producer and consumer countries. For a further description and analysis of the International Energy Agency and its workings, compare Willrich and Conant (3).

[3]An emergency is defined as an actual or potential shortfall in oil supplies either by the group as a whole or by any one of the member countries. If the shortfall affects the group, then each member country is to reduce its oil consumption by 7 percent, and the total oil available will then be shared among member countries according to a predetermined formula based on the drawdown of the oil stockpile, actual oil imports, and production.

But the major part of these countries' energy programs is of course implemented within the domestic context. In order to give some idea of the nature and extent of energy policy reactions among our selected countries, we will offer a brief survey of developments, particularly with regard to conservation policies in each country since the 1973-74 embargo and price rise. The (relatively) resource-rich countries will be considered first.

United States

The United States looms large in the world oil picture--as a major producer, a major consumer, and a major importer. The United States accounts for 44 percent of total OECD oil consumption and for 28 percent of total OECD oil imports. The U.S. share of OECD oil imports has risen sharply from its 15 percent share in 1970.

There has clearly been an important change in perceptions about energy policy in the United States since 1973, particularly regarding the role of the federal government in this policy. This is reflected in the establishment of a unified Department of Energy in place of the many separate agencies that had existed before. It is also reflected in the leadership which the United States assumed in setting up the International Energy Agency, and the wide-ranging energy programs which were submitted to Congress by both Presidents Ford and Carter.

As is well known, legislative achievement fell far short of the presidential intentions. After a lengthy discussion, the following provisions have emerged:

- A petroleum stockpile has been authorized.
- Mandatory fuel economy standards for automobiles have been adopted.

- Appliances must be labeled for energy efficiency.
- Gradual decontrol of natural gas prices has been specified.
- State public service commissions must consider the use of electricity rates encouraging conservation.
- Energy conservation and use of solar devices are to be subsidized.
- Taxes on gas-guzzling automobiles are to be increased.
- Major boiler users are to be encouraged to switch from oil to coal.
- Minimum insulation standards for houses have been laid down.

On the central issue of decontrol of petroleum prices, little was done until 1978, when President Carter announced the phasing-out of controlled oil prices. This is to be completed by late 1980.

Canada

The Canadian government's reaction to the oil crisis was to proclaim a policy of self-sufficiency--at first sight, this was not an overambitious policy, for Canada has been historically self-sufficient in energy. More particularly the oil price rise focused attention on Canada's oil reserves and their subsequent development. Although domestic crude oil prices were allowed to rise, higher federal and provincial government "taxes" absorbed much of the profit, and it was not until tax burdens were reduced (in a series of steps from late 1974 to 1976) that *de facto* incentives for exploration were increased. The Canadian government is committed in principle to move domestic oil prices toward international levels and natural gas prices to a competitive relationship with oil. But a major constraint on their doing so is their determination not to exceed the U.S. average price.

The government set up a national oil company, Petrocanada, with authority to act in all aspects of petroleum R&D exploration, production, distribution, refining, and marketing. Priority appears to have been assigned to exploration in frontier areas and the development of domestic (as opposed to foreign) reserves.

On the conservation side, specific provisions have been made for establishing fuel economy standards for automobiles. Higher gas taxes have encountered opposition, and indeed may have been a significant factor in the fall of the short-lived conservative government of Prime Minister Clark in 1980.

The Netherlands

Because of its indigenous supplies of natural gas, the Netherlands has, compared with most of our other countries, low rates of import dependence (about 8 percent in 1978). After years of an aggressive development and export policy, the Netherlands government now anticipates a reduction in output, and plans not to renew export contracts when they expire. The reduction in gas production is to be compensated for within a limited period by an increase in oil imports and by imports of liquefied natural gas (LNG).

With regard to conservation, gas prices are being brought gradually into parity with petroleum prices, new building codes have been introduced, and district heating has been stimulated by financial incentives. In industry, subsidies are available for investment in energy-saving equipment. Lower speed limits apply to transport.

The United Kingdom

The United Kingdom is in the fortunate position of having North Sea oil come onstream at the same time the international oil market has been disrupted. These new energy resources, combined with the already established coal industry, mean that the United Kingdom will soon be self-sufficient in energy. Even now, it is only 34 percent import dependent--a relatively low level for a European country.

With regard to conservation, the emphasis is on voluntary programs: the government's role is restricted to providing information about energy efficiencies. Building codes have been tightened, and energy-conservation investments in public buildings (including the large public housing sector) will be made. The amount of oil used to generate electricity has been reduced.

West Germany

In Germany, where domestic coal production results in somewhat lower rates of energy-import dependency (61 percent) than those of most other European countries, the immediate reaction to the events of 1973-74 was more measured. The original 1973 Energy Policy Program relied entirely on supply strategies, including higher domestic production and substitution of other fuels for oil. It was thought initially that this strategy would be adequate to deal with the new situation. Opposition to rapid expansion of nuclear facilities has, however, led to recent greater interest in conservation measures. The tax on heating oil has been doubled, the proceeds to be used to subsidize insulation and the installation of solar collectors and heat pumps; stricter standards for insulation of new buildings have been mandated; domestic appliances are labeled for energy efficiency; industrial energy-saving investments are to benefit from tax credits; and a program for district heating has been introduced. The construction of oil- and gas-fired power plants has been prohibited (5).

Others

The remaining countries--France, Italy, Japan, and Sweden--are poor in energy resources.

France

France is the only country among the nine analyzed here that is not a member of the International Energy Agency. Perhaps because of its relatively high level of energy-import dependence, France initiated, soon after the 1973-74 price rise, an active energy program which included the requirement that oil imports be held below a fixed-value ceiling. More recently, the drive toward conservation has been slowed because of the Council of Ministries' rejection of a series of conservation measures. On the other hand, the large-scale nuclear program, including a fourfold increase in nuclear capacity between 1979 and 1985, and a strong commitment to the breeder, enjoys a wider measure of support in France than in many other countries.

Italy

Italy is relatively poor in energy resources and, like France, has high rates of import dependency (80 percent). Although it plans to increase oil imports as part of a general policy of increasing economic growth and reducing employment, tax rates on petroleum products have been increased and natural gas used in the residential sector has been recently taxed to bring its price into parity with light fuel oil. Subsidies to electricity tariffs are to be reduced. In addition to offering fiscal incentives to conservation, standards have been imposed for heating installations and room temperatures for all buildings, and an insulation code for new buildings has been enacted. And speed limits have been set for cars.

On the supply side, the nuclear program is to be rapidly developed, and contracts have been made with Algeria for the delivery of 12 billion cubic meters of natural gas per annum for twenty years.

Japan

Japan is perhaps the most extreme example of resource penury. Energy imports account for over 90 percent of its total consumption. Opportunities for substitution of domestic production (nuclear generation of electricity) or even other imports (coal and LNG instead of oil) are limited, because of strong public opposition to siting of such facilities.

So far, little attention has been paid to conservation and, indeed, the low levels of energy consumption in both the residential and transportation sectors suggest that further reductions, especially in the face of rising incomes, may be difficult. As in some of the other countries, however, the use of oil in electricity generation has been limited.

Sweden

Sweden is saved from very high import dependency by its hydroelectric resources. But as Sweden has virtually no indigenous fossil fuels and depends for its supplies on imported oil, energy policy has excited a good deal of concern over the past years.

An energy policy program was announced in 1975 which, at first sight, marked a clear break with the past (6, 7). The primary, and most highly publicized, feature of the program was its determination to reduce the growth rate of net energy consumption from the 4.5 percent annual rate of the postwar period to an average of 2 percent per year up to 1985, and from then to keep consumption at a constant level, that is, zero energy growth. This was, and still is, the most radical declaration of intention made by any of the industrialized countries. In practice, however, this concept has been modified by the proviso that zero energy growth would only be attained provided that it was not in conflict with other economic and social goals.

The contraction in energy growth rates was to be achieved by expanding district heating systems and the use of waste heat, energy conservation in homes and buildings, industrial energy conservation, educational and information activities on energy conservation, and increasing taxes on gasoline, bottled gas, oil, coke, and electricity.

In order to assure that adequate supplies are available, some limited expansion of hydro and nuclear facilities was envisaged, combined electricity and heat production units were encouraged, and an active oil policy was adopted This includes financial support for the prospecting and acquisition of oil deposits or production facilities abroad, the formation of a new oil trading company, and increases in reserve stocks of crude oil and petroleum products.

Since that time, of course, the issue of nuclear expansion has moved center stage and has been the central issue in two general elections. In 1980 a referendum on the future of nuclear energy was held. The results showed support for continuing the nuclear program, albeit a somewhat limited program.

The results of the referendum are, however, advisory, without the force of law. Both conservation and supply programs were underpinned by a comprehensive three-year research and development program with a total expenditure of 366 million Kronor, including energy use in industry heating and ventilation of buildings; recycling of energy intensive goods; energy production, including district heating, methanol, integration of industry in energy production systems, and new conversion processes.

From this brief survey of policy measures introduced after 1973-74, it can be concluded that there were indeed policy reactions to the new situation. The greatest change, however, appears to have been in perception rather than in action. Thus, many countries issued energy plans or statements of energy policy and intent. The interest of these documents is twofold. First, the energy sector was treated as a whole instead of each single fuel supply being isolated, as had been done in the past. And, second, all plans paid attention to demand, either by specifying their intention of reducing the rate of increase in energy consumption, or at a minimum, by including lower energy consumption scenarios in forecasting exercises.

Countries without domestic resources emphasized reduction of total energy consumption (which in these cases consists largely of imported oil). Countries with domestic energy resources emphasized reducing oil consumption either through conservation measures or thourgh substituting other fuels for oil. The concern of the industrialized countries about the future of oil is reflected in their repeatedly expressed intentions (in summit meetings and in the Ministerial meetings of the International Energy Agency) to keep oil imports in 1985 at or below current levels.

Although these declarations of interest are indicative of a changing perception of energy policy, including consciousness-raising about conservation, they do not, of course, in themselves guarantee that these intentions will be fulfilled. Implementation has proved difficult in many countries--particularly the imposition of speed limits--because of political opposition and, therefore, it has been rather limited in nature. Action to increase domestic production--especially of nuclear power--has also been hampered in many countries by strong opposition, and there is some doubt whether coal production can be raised to hoped-for levels.

On the demand side, most countries--whether possessing an interventionist or a market orientation--have relied primarily on the increase in petroleum product prices stemming from the rise in crude oil to moderate demand, although some moves have also been made to tighten building codes and to label the energy efficiency of appliances. Several countries seek to limit the use of oil in electricity generation.

On balance, though much more attention is now paid to conservation, action still appears to be largely supply oriented. Conservation, for example, accounts for under 10 percent of total energy R&D budgets, with the single exception of Sweden (39 percent), whereas the major supply sectors (oil, gas, coal, and nuclear) account for almost 70 percent (8). Nuclear energy alone accounts for one-half.

Are the changes instituted so far sufficient to fulfill stated objectives? In the opinion of the IEA which follows closely and on a continuing basis the energy policies of their member countries, they are not. In the last review of energy policies (8), they recommended stepped-up activity on all fronts. With regard to conservation they recommended:

- Prompt enactment of comprehensive energy legislation in several countries, especially the proposed National Energy Plan in the United States

- Continued and expanded emphasis on conservation measures, particularly retrofitting incentives, increased automobile efficiency, less energy-intensive industrial processes, and increased use of waste heat, district heating, and combined production of heat and electricity

- Energy price structure designed to encourage conservation and production of alternative energy sources

Failing further policy initiatives along these lines, the IEA foresees imports from outside the area ranging between 31 and 36 million barrels per day (mbd) in the mid-1980s, substantially above the targeted 25 mbd.

There are, of course, understandable reasons for the slow and inadequate policy responses. Given the uncertainty about future energy prices, especially for the first year or so after the rise, when it was by no means clear that prices could be maintained at that level, governments and individuals were reluctant to make expensive commitments in energy-producing or consuming projects with a long life. Second, in many cases, energy policy in the sense that it is now employed did not have the institutional framework needed for its implementation. New energy ministries and agencies had to be set up, and in some countries, new powers had to be sought. Third, as already indicated, energy consumption fell markedly relative to the GDP in the OECD area in 1974 and 1975, as did oil imports into Europe, perhaps inducing a sense of false security. It is not clear how far such reductions were caused primarily by higher oil prices, industrial recession, or government policies.

These comments should not be interpreted as an indictment of conservation policies. Their apparent lack of effect to date can be attributed to the fact that many of them--tighter building codes, fuel-efficiency standards--will take many years to affect energy consumption substantially. The difficulty is, however, that many of our energy objectives--such as reducing oil imports--apply to the short to medium term.

In summary, the events of 1973-74 stimulated a changing perception of energy policy in the direction of a more coherent and comprehensive approach to the energy sector. Policy responses still, however, tend to give more emphasis to supply rather than to conservation questions. There seems little reason to argue with the IEA's assessment of the current level of policy response as being inadequate to fulfill stated targets within the given time frame. While the reluctance to act more strongly so far is understandable, there will be need for further action if stated policy objectives are to be achieved.

References

1. Organisation for Economic Co-operation and Development; Energy Policy--Problems and Opportunities (Paris; OECD, 1966).

2. Joy Dunkerley, "The Future for Coal in Western Europe," Resources Policy (September 1978).

3. M. Willrich and M. A. Conant, "The International Energy Agency: An Interpretation and Assessment," The American Journal of International Law vol. 71, 1977.

4. "OECD's New International Energy Agency," OECD Observer no. 73 (January-February) 1975.

5. Federal Ministry of Economics, Energy Policy Program for the Federal Republic of Germany, second revision (December 14, 1977).

6. M. Lonnroth, P. Skeen, and T. B. Johansson, Energy in Transition (Stockholm, Secretariat for Future Studies, Swedish Institute, 1977).

7. Leon N. Lindberg, The Energy Syndrome (Lexington, Mass., Lexington Books, 1977).

8. International Energy Agency/Organisation for Economic Co-operation and Development, Energy Policies and Programs of IEA Countries, 1977 Review (Paris, OECD, 1978).

Chapter 9

FINDINGS OF THE STUDY

Our study has revealed both diversity and similarity in patterns of energy consumption among countries and over time. Diversity is illustrated by the fact that the amount of energy consumed relative to a given amount of the GDP--the energy/GDP or output ratio--varies substantially among countries and over time within the same country. Within this diversity, however, there are indications, particularly in the 1960-73 period, that patterns of energy consumption within the countries of Europe and Japan are approaching those of the United States. For a given level of income, the amount of energy consumed in industrial and nonenergy uses is now very similar in both the United States and the other countries. And energy consumed by residential and transportation uses, while still substantially below the U.S. level, was rapidly catching up, at least until 1973.

During this period of steady economic growth and declining energy prices, a broad picture of energy consumption patterns emerges. The United States, the richest country, appeared to have reached a plateau, with its energy consumption roughly keeping pace with its economic growth and with relatively little change being seen in the sectoral distribution of energy demand. On the other hand, energy consumption in the other industrial countries, particularly if expressed in useful terms, tended to rise faster than national output, and there were radical changes in the sectoral distribution of energy consumption, with the residential-commercial and transportation sectors taking an increasing share. If this story had ended at 1973, it would have been tempting to conclude that energy consumption in Western Europe and Japan would have continued to rise rapidly, especially in those sectors where consumption levels were still substantially below those of the United States. This trend might have been expected to continue until levels of affluence and amenity similar to those in the United States were attained. At that point, rises in energy consumption might have moderated, just keeping pace with rises in economic output. In other words, the other countries would have reached the plateau which appeared to typify U.S. experience in this period.

Developments in recent years have changed all of these reasonable prognostications. In the two years after the 1973-74 oil price increases, the energy/GDP ratios of all countries fell. Factors other than price increases--industrial recession, and mild winters--also contributed to this general decline. While it is difficult to say which predominated, their combined effect was sufficiently strong to override the considerable variations in experience noted among these countries prior to 1973.

Structure of the Energy Supply System

This study is directed to investigating the reasons behind these variations in energy use, both convergence and diversity. We begin by analyzing the effects of changes in the structure of energy supply on differences in energy consumption. Two major developments in fuel supply systems were taking place. The first was the widespread and rapid substitution of oil for coal in the European countries and Japan, but not in the United States where the substitution cycle had taken place a generation earlier. As oil was used in more thermally efficient ways than the coal it replaced, the amount of useful energy obtained was greater than would have been obtained by an equivalent amount of coal.

A second development was the divergence in energy losses (mainly heat losses incurred in generating electricity from fossil fuels) experienced between the United States and the other countries. These losses have tended to increase in the United States as a share of total gross energy consumption and to fall in the other countries.

The combined effect of these two developments was to increase the amount of useful energy or energy services achieved from a given energy input in the countries of Western Europe and Japan. In other words, the energy supply of these countries was becoming more efficient. This meant that energy consumption relative to the GDP, expressed in useful terms, was rising more rapidly than might be supposed from an inspection of gross energy consumption data. It also meant, as there was little change in the efficiency of U.S. energy supply, that energy consumption in Western Europe and Japan was approaching U.S. levels more rapidly. Differences in the structure of the fuel supply, therefore, account for part of the difference in energy consumption relative to the GDP over time, and for part of the seemingly lower energy consumption relative to output in the European countries and Japan.

The Effect of Differences in Composition of Economic Output

The amount of energy consumed relative to the GDP is also affected by the composition of national output. Some activities, such as steel production, are inherently energy-intensive so that a country with a large steel industry is sure to consume more energy relative to output than will a country whose output is concentrated on agriculture, for example. Although all our countries have common characteristics, such as a broad industrial base and relatively high standards of living, there are nonetheless important differences among them which would be expected to influence the amount of energy being used. A previous study (1) found that almost one-half of the higher U.S. energy consumption relative to the GDP was caused by differences in the composition of economic activities, including long-standing characteristics of the transportation and residential sectors. We therefore investigated how far changes in energy/GDP ratios over a period of time, rather than among countries, were caused by changes in composition of economic output and activities.

Our analysis, because of formidable data requirements, was limited to two countries, the United States and Sweden and, in the case of Sweden, to a limited period, from the mid-1960s to the mid-1970s. It indicated that changes in compositional characteristics, such as the growth of the electricity-generating sector and the petrochemical industry, did contribute significantly to changes in energy/GDP ratios over time. Another factor of importance to all countries was the generally declining amounts of energy used per unit of industrial output in this period.

Energy Prices, Income, and Energy Consumption

Finally, the role of energy prices in explaining differences in energy consumption both among countries and over time was assessed. U.S. energy prices were consistently lower and more stable than those of other countries. Energy prices in real terms declined until 1973, but the decline in European countries tended to be steeper than that in the United States. In 1974 energy prices increased sharply (about 20 to 30 percent in real terms) in all countries, but up to early 1978 they tended to remain stable in real terms or to decline, particularly in the household and transportation sectors.

Our analysis suggests that higher energy prices have considerable impact on energy consumption. More formally, we found that a given increase in energy prices in these countries, say, about 10 percent, was associated with a less than proportional, but still significant, reduction in energy consumption, say, about 5 percent. This relationship is assumed to apply over the long term rather than the short. In the short term, the existence of a large stock of energy-using equipment precludes radical changes in energy-consuming habits. There is reason to suggest that the responsiveness of energy consumption to energy prices varies considerably from sector to sector, being particularly high in transportation.

While energy consumption appears, therefore, to be moderately responsive to energy price changes, the relationship between increases in income and energy consumption appears to be stronger. Thus, a given increase in the GDP was associated in these countries with a somewhat greater increase in consumption of useful energy or energy services.

This means that rising incomes played a major part in the rising useful energy/GDP ratios of the European countries during 1960-73. Declining real prices also contributed, but the generally low values of the price elasticities, and, in many cases, their uncertain significance, imply that, over time, income was the major force behind changes in energy/GDP ratios. After 1973 there is evidence of the greater influence of price, but here again it is difficult to be specific as the fall in energy/GDP ratios which took place after this year also owed something to mild winters and the industrial recession. Differences among countries in energy/GDP ratios, on the other hand, are largely caused by different levels of energy prices.

The convergence between the U.S. ratio and that of other countries over time shows the influence of the higher income elasticity in European countries. The considerable gap that remains is caused by higher European energy prices.

Influence of Energy Policies

In the period up until 1973 we have seen that policies of economic management in all countries were conducive to a rapid and sustained increase in total and per capita energy consumption. It is not clear, however, whether such policies would necessarily result in an increasing or decreasing energy/GDP ratio. Insofar as these economic policies led to the rapid development of energy-intensive industries (as in the case of the Netherlands and Italy), then energy/GDP ratios tended to rise. In other cases, as in Japan, the rapid growth rate leading to a quick turnover in capital stock--usually in an energy-saving direction--tended to reduce the energy/GDP ratio.

These trends were modified by energy-specific policies. During this period, most countries, through their access to Middle Eastern oil, were able to provide cheap, plentiful, supplies of energy to fuel their rapid economic growth. Consequently, declining relative prices, particularly in Europe, contributed to the generally increasing energy/GDP ratios. Noteworthy in the case of the Netherlands was the decision to develop the Groningen gas field rapidly at very advantageous prices. Certainly, this was the main factor leading to the very rapid rise in that country's energy/GDP ratio. On the other hand, those countries who protected their domestic energy industries, therefore moderating the decline in energy prices, tended to have steady or declining energy/GDP ratios.

The importance of energy pricing and conservation policies becomes clearer after 1973. Following the sharp rise in energy prices in 1974, energy consumption relative to the GDP in all our countries declines. While several factors entered into this decline, higher prices and new conservation policies appear to have contributed. Thus, according to the International Energy Agency (2), "By and large, countries that have experienced large increases in energy prices and that have strong conservation programmes have been more successful in reducing their energy use per unit of economic growth. Whereas countries that have experienced small energy price increases and have introduced few conservation measures generally have been less successful in reducing their energy use per unit of economic growth. Thus both energy pricing and conservation measures seem to have had a significant effect in improving energy efficiencies."

While all evidence appears to point to the effectiveness of higher prices, reinforced by supplementary conservation measures, in moderating energy consumption, there is general agreement that the level of conservation effort has not so far been adequate to achieve wider policy aims, in particular, a significant reduction in oil import dependency within the next few years.

Implications for the Future

In summary, this report makes it clear that there is a wide range of energ consumption associated with a given value of output and that the amount of energy consumed is not fixed in all circumstances and times. To this extent, it gives grounds for optimism in assessing energy conservation possibilities. But, as we have also seen, some of these differences are caused by such factors as differences in the composition of economic output, in the structure of fuel supply, in the vintage of energy-using equipment, in geography and tastes, and in energy prices, which may be difficult to change radically, particularly in the short run. The prevailing economic climate, which determines how easy it is for industries to pass on higher energy prices to their customers, will also affect the amount of energy consumed. At a minimum, this means that, taken by themselves, energy/GDP ratios are at best a partial indicator of the energy conservation potential among countries or of progress in energy conservation over time.

More generally for the future of energy consumption, this analysis indicates a very strong underlying demand for energy services, especially in the European countries and Japan. In particular, we found, during the 1960s at least, a tendency for consumption of energy services in these countries to outstrip growth in the GDP, leading to rising useful energy/GDP ratios.

In the two years following the 1973-74 oil price increases, energy/GDP ratios fell. But factors other than prices--industrial recession and, for 1976 at least, exceptionally mild winters--also contributed to this fall. The question is, what will happen to energy consumption when economic growth resumes? On the basis of this study we can anticipate a continued strong demand for energy services, but a demand now restrained by rising real energy prices. As an illustration of rough orders of magnitude involved: if these countries were to achieve a 3 percent average increase in GDP per capita in the years to come, the demand for energy services (given an income elasticity of about 1.2) would therefore be expected to rise by 3.6 percent. In order to keep the rise in energy consumption to a level of 2 percent a year, which is the aim of many countries, relative energy prices (given a price coefficient of about 0.5) would need to rise by about 3 percent a year.

Given the sharp increase in oil prices that occurred in 1979 and expectations of more to come, rises of this magnitude in energy prices to consumers might appear to be a foregone conclusion. But, as we have seen, price rises in crude petroleum translate into much smaller energy prices paid by the consumers. Despite the four- or fivefold increase in petroleum prices between 1973 and 1978, average real energy prices to all consumers rose by only about 20 percent. Real prices of gasoline and household fuel and power in several OECD countries were in fact no higher at the beginning of 1978 than it was in 1972. This experience suggests that if other factors (such as strong rates of inflation or depreciation of the dollar) intervene, the rise in real prices of energy to purchasers may not be sufficient to trigger hoped-for conservation efforts. Further, the price elasticity given here is assumed to reflect a long-term rather than a shorter-term adjustment. In this case, an increase in prices may have the effect of reducing energy consumption, but not within the desired time frame.

All in all, it would be unwise to count on a steady rise in real energy prices, stemming from the higher prices of imported oil, being sufficient in and of itself to modify the apparently strong demand for energy services to a level consistent with stated policy goals, such as the limitation of OECD oil imports to 26 million barrels per day in 1985. In any event, given the uncertainty which surrounds the behavior of energy consumption, especially since 1974, it is only prudent to consider other policy measures.

The analysis of the sectoral components of energy consumption and the detail provided by the case studies of the United States and Sweden further reinforce this conclusion. Within total energy demand, the demand for energy services connected with passenger transport and residential-commercial uses (primarily heating) was particularly strong and might, in the European countries, be expected to increase further as income increases to approach U.S. levels. Furthermore, an admittedly limited analysis of structural and intensity factors suggests that there was a strong structural effect behind the rise in energy consumption. In this event, it might be necessary not only to supplement price incentives by other conservation mechanisms, but perhaps also to take a wider sectoral view of energy management than has been taken so far.

REFERENCES

1. J. Darmstadter, J. Dunkerley, and J. Alterman, *How Industrial Societies Use Energy*: A Comparative Analysis (Baltimore, Md., Johns Hopkins University Press for Resources for the Future, 1977).

2. International Energy Agency/Organisation for Economic Co-operation and Development, *Energy Conservation in the International Energy Agency, 1978 Review* (Paris, OECD, 1979).

Appendix A

AGGREGATE AND SECTORAL ENERGY INTENSITIES FOR SELECTED COUNTRIES

Table A-1. Aggregate Energy Consumption, Per Unit of Gross Domestic Product

Country	1960	1964-66	1970	1973	1976
	Panel A (toe/million $ U.S.)				
United States	1,408	1,351	1,453	1,401	1,360
Canada	1,728	1,651	1,677	1,742	1,632
France	823	814	775	808	710
Germany	1,011	1,002	1,015	1,026	973
Italy	631	730	865	876	855
The Netherlands	823	940	1,088	1,207	1,194
United Kingdom	1,216	1,182	1,171	1,119	1,010
Sweden	1,050	1,074	1,114	1,121	1,123
Japan	867	852	908	848	800
Average excluding United States and Canada	917	942	991	1,001	952
	Panel B (U.S. = 100)				
Canada	123	122	115	124	120
France	58	60	53	58	52
Germany	72	74	70	73	72
Italy	45	54	60	63	63
The Netherlands	58	70	75	86	88
United Kingdom	86	87	81	80	74
Sweden	75	79	77	80	83
Japan	62	63	62	61	59
Average excluding United States and Canada	65	70	68	71	70

Source: International Energy Agency/Organisation for Economic Co-operation and Development. Energy Balances of OECD Countries, various issues (Paris, OECD).

Table A-2. Energy Losses, Per Unit of Gross Domestic Product

Country	1960	1964-66	1970	1973	1976
		Panel A (toe/million $ GDP)			
United States	283	297	321	351	357
Canada	601	519	502	604	531
France	245	216	177	164	165
Germany	303	268	253	239	253
Italy	173	162	196	209	186
The Netherlands	252	243	253	234	208
United Kingdom	337	366	365	348	306
Sweden	328	318	270	264	292
Japan	306	249	210	237	210
Average excluding United States and Canada	277	60	246	243	231
		Panel B (U.S. = 100)			
Canada	212	175	156	172	149
France	87	73	55	47	46
Germany	107	90	79	69	71
Italy	60	55	61	60	52
The Netherlands	89	82	79	67	58
United Kingdom	119	123	114	99	86
Sweden	116	107	84	75	82
Japan	108	84	65	68	59
Average excluding United States and Canada	98	88	77	69	65

Source: International Energy Agency/Organisation for Economic Co-operation and Development. Energy Balances of OECD Countries, various issues (Paris, OECD).

Table A-3. Industrial Energy Consumption, Per Unit of Gross Domestic Product

Country	1960	1964-66	1970	1973	1976
	Panel A (toe/million $ U.S.)				
United States	378	363	371	297	239
Canada	378	365	377	351	341
France	265	259	257	219	189
Germany	336	305	291	287	244
Italy	219	258	287	254	233
The Netherlands	214	227	263	274	255
United Kingdom	346	321	310	290	248
Sweden	336	314	338	348	317
Japan	335	307	345	288	280
Average excluding United States and Canada	293	284	299	280	252
	Panel B (U.S. = 100)				
Canada	100	101	102	118	143
France	70	71	69	74	79
Germany	89	84	78	97	102
Italy	58	71	77	86	99
The Netherlands	57	63	71	92	106
United Kingdom	92	88	84	98	104
Sweden	89	87	91	117	133
Japan	89	85	93	97	119
Average excluding United States and Canada	78	78	80	94	106

Source: International Energy Agency/Organisation for Economic Co-operation and Development. Energy Balances of OECD Countries, various issues (Paris, OECD).

Table A-4. Energy Consumption for Non-Energy Uses Per Unit of Gross Domestic Product

Country	1960	1964-66	1970	1973	1976
Panel A (toe/million $ U.S.)					
United States	52	43	59	79	73
Canada	54	59	65	67	56
France	22	26	43	63	40
Germany	22	47	63	69	59
Italy	25	57	71	70	95
The Netherlands	8	49	93	155	152
United Kingdom	32	42	56	58	52
Sweden	31	24	43	55	98
Japan	16	72	100	58	74
Average excluding United States and Canada	22	45	67	75	81
Panel B (U.S. = 100)					
Canada	104	137	110	85	77
France	42	60	73	80	55
Germany	42	109	107	87	81
Italy	48	133	120	89	132
The Netherlands	15	114	158	196	208
United Kingdom	62	98	95	73	71
Sweden	60	56	73	70	134
Japan	31	167	169	73	100
Average excluding United States and Canada	43	105	114	95	112

Source: International Energy Agency/Organisation for Economic Co-operation and Development. Energy Balances of OECD Countries, various issues (Paris, OECD).

Table A-5. Residential-Commercial Energy Consumption Per Unit of Gross Domestic Product

Country	1960	1964-66	1970	1973	1976
	Panel A (toe/million $ U.S.)				
United States	370	342	379	348	360
Canada	401	423	436	413	390
France	185	204	189	241	198
Germany	239	266	288	303	288
Italy	129	140	188	212	227
The Netherlands	244	303	343	398	414
United Kingdom	351	320	293	269	255
Sweden	247	293	338	336	333
Japan	102	123	150	165	139
Average excluding United States and Canada	214	236	256	275	265
	Panel B (U.S. = 100)				
Canada	108	124	115	119	108
France Germany	50	60	50	69	55
Germany	65	78	76	87	80
Italy	35	40	50	61	64
The Netherlands	66	89	91	115	115
United Kingdom	95	94	77	77	71
Sweden	67	86	89	97	93
Japan	28	36	40	47	39
Average excluding					75
Canada	58	69	68	79	75

Source: International Energy Agency/Organisation for Economic Co-operation and Development. Energy Balances of OECD Countries, various issues (Paris, OECD).

Table A-6. Transportation Energy Consumption, Per Unit of Gross Domestic Product

Country	1960	mid-60s	1970	1973	1976
Panel A (toe/million $ GDP)					
United States	325	305	323	326	332
Canada	288	284	298	307	314
France	106	108	109	120	119
Germany	110	114	120	127	129
Italy	86	114	124	130	115
The Netherlands	105	122	137	146	165
United Kingdom	151	141	148	154	149
Sweden	112	120	124	124	127
Japan	108	103	103	100	97
Average excluding United States and Canada	111	117	124	129	129
Panel B (U.S. = 100)					
Canada	89	93	92	94	95
France	33	35	34	37	36
Germany	34	37	37	39	39
Italy	26	37	38	40	35
The Netherlands	32	40	42	45	50
United Kingdom	46	46	46	47	45
Sweden	34	39	38	38	38
Japan	33	34	32	31	30
Average excluding United States and Canada	34	38	38	40	38

Source: International Energy Agency/Organisation for Economic Co-operation and Development. Energy Balances of OECD Countries, various issues (Paris, OECD).

Table A-7. Changes in Aggregate and Sectoral Energy Intensities (toe/million $ GDP)

Country/sector	1960-1973	1973-1976
United States	-7	-41
Total	-7	-41
Energy losses	68	6
Industry	-81	-58
Transportation	1	6
Residential-commercial	-22	12
Nonenergy	27	-6
France		
Total	-15	-98
Energy losses	-81	1
Industry	-46	-30
Transportation	14	-1
Residential-commercial	56	-43
Nonenergy	41	-23
Germany, F. R.		
Total	15	-53
Energy losses	-64	14
Industry	-49	-43
Transportation	17	2
Residential-commercial	64	-15
Nonenergy	47	-10
Italy		
Total	245	-22
Energy losses	36	-23
Industry	35	-21
Transportation	44	-15
Residential-commercial	83	15
Nonenergy	46	25
The Netherlands		
Total	384	-13
Energy losses	-18	-26
Industry	60	-19
Transportation	41	19
Residential-commercial	154	16
Nonenergy	147	-3

(continued on next page)

Table A-7 (continued)

Country/sector	1960-1973	1973-76
United Kingdom		
Total	-97	-109
Energy losses	11	-42
Industry	-56	-42
Transportation	3	-5
Residential-commercial	-32	-14
Nonenergy	26	-6
Japan		
Total	-19	-48
Energy losses	-69	-27
Industrial	-47	-8
Transportation	-8	-3
Residential-commercial	63	-26
Nonenergy	42	16

Source: Appendix A tables.

Appendix B

ADJUSTMENT OF SECTORAL ENERGY DATA TO TAKE INTO ACCOUNT VARYING THERMAL EFFICIENCIES OF FUELS

The sectoral energy consumption data given in OECD, Energy Balances of OECD Countries, net of heat losses and consumption by the energy industries, were multiplied by the following thermal efficiencies to arrive at "useful" energy consumption.

Fuels	Sectors		
	Residential-commercial	Transport	Industry, except energy
Solid	0.20	0.044	0.70
Liquid	0.60	0.22	0.80
Gas	0.70	0.22	0.85
Electricity	0.95	0.40	0.99

Source: W. D. Nordhaus, editor, Proceedings of the Workshop on Energy Demand. International Institute for Applied Systems Analysis. (Laxenburg, Austria, IIASA, 1975), p. 527, based in turn on H. C. Hottel and J. B. Howard, New Energy Technology: Some Facts and Assessments (Cambridge, Mass., MIT Press, 1971).

Joy Dunkerley

Joy Dunkerley has devoted the past ten years to research and writing on energy conservation. She was a staff economist on the Ford Foundation Energy Policy Project which produced the first comprehensive set of energy consumption scenarios for the United States. As a fellow, then senior fellow, of Resources for the Future, she has completed three studies on the energy conservation lessons to be drawn from comparing U.S. energy consumption patterns with those of other industrial countries. She is currently engaged in research on energy conservation possibilities in developing countries.